Harmonic Maps and Minimal Immersions Through Representation Theory

PERSPECTIVES IN MATHEMATICS, Vol. 12

J. Coates and S. Helgason, editors

Harmonic Maps and Minimal Immersions Through Representation Theory

Gabor Toth

Department of Mathematical Sciences
Rutgers University
Camden, New Jersey

ACADEMIC PRESS, INC.
Harcourt Brace Jovanovich, Publishers

Boston San Diego New York
Berkeley London Sydney
Tokyo Toronto

ACADEMIC PRESS, INC.
1250 Sixth Avenue, San Diego, CA 92101

United Kingdom Edition published by
ACADEMIC PRESS LIMITED
24-28 Oval Road, London NW1 7DX

Library of Congress Catalog Card Number: 89-46479
ISBN: 0-12-696075-5

Printed in the United States of America
90 91 92 93   9 8 7 6 5 4 3 2 1

Dedicated to

James Eells Jr.

CONTENTS

PREFACE

One of the oldest examples of a minimal surface carries the name of Veronese. Topologically it is a real projective plane $\mathbf{R}P^2$ imbedded minimally into S^4. Analytically it is defined as the image of the (polynomial) Veronese map

$$f_{\lambda_2} : S^2 \to S^4$$

given by

$$f_{\lambda_2}(x, y, z) = \left(\frac{1}{\sqrt{6}}(2x^2 - y^2 - z^2), \frac{1}{\sqrt{6}}(2y^2 - x^2 - z^2), \frac{1}{\sqrt{6}}(2z^2 - x^2 - y^2), \right.$$
$$\left. \frac{1}{\sqrt{8}}xy, \frac{1}{\sqrt{8}}xz, \frac{1}{\sqrt{8}}yz \right), \quad (x, y, z) \in S^2 \subset \mathbf{R}^3.$$

At the first sight f_{λ_2} seems to map into $S^5 \subset \mathbf{R}^6$. Notice, however, that the first three components add up to zero so that the image of f_{λ_2} is further contained in the corresponding hyperplane of $\mathbf{R}^6$ that cuts out an S^4 from S^5. (Notice also that, being quadratic, f_{λ_2} factors through the antipodal map of S^2.)

For our purposes, the most important feature of f_{λ_2} is that its components are harmonic homogeneous polynomials on $\mathbf{R}^3$. Resticting to $S^2 \subset \mathbf{R}^3$ they become spherical harmonics (of order 2), i.e. eigenfunctions of the Laplace-Beltrami operator $\triangle^{S^2}$ (corresponding to the second eigenvalue λ_2 that has been prematurely affixed to f). A more elegant (and thereby less revealing) way to define f_{λ_2} is to consider the (5-dimensional) eigenspace $\mathcal{H}^2_{S^2}(\mathbf{R})$ of quadratic spherical harmonics on S^2 and introducing f_{λ_2} as a map whose components comprise an orthonormal basis in $\mathcal{H}^2_{S^2}(\mathbf{R})$ with respect to a suitably normalized L^2-scalar product given by the standard Riemann measure on S^2. Easy integration shows that the two definitions are equivalent. However abstract, the second definition tells us that f_{λ_2} has as many symmetries as possible. Namely, it is equivariant with respect to the homomorphism $\rho_\lambda : SO(3) \to SO(5)$ that is just the (irreducible orthogonal) $SO(3)$-module structure on $\mathcal{H}^2_{S^2}(\mathbf{R})$ given by $a \in SO(3)$ acting as $a \cdot \mu = \mu \circ a^{-1}$, $\mu \in \mathcal{H}^2_{S^2}(\mathbf{R})$. Hence we can look at the Veronese surface as an orbit of the action of $SO(3)$ on the unit sphere S^4 of $\mathcal{H}^2_{S^2}(\mathbf{R})$ which, as a little inspection shows, is one of the two singular orbits. (In fact, the orbits on S^4 form another geometric object of interest: a homogeneous (and hence isoparametric) family of hypersurfaces (of degree 3) but this is not our main concern here.)

What we can learn about this example can be substantially generalized. As the thorough treatment will be given in the main text here we only replace S^2 by S^3 to get

$$f_{\lambda_2} : S^3 \to S^8$$

given by

$$f_{\lambda_2}(x, y, u, v) = \sqrt{\frac{2}{3}} \left(\frac{1}{\sqrt{2}}(x^2 + y^2 - u^2 - v^2), x^2 - y^2, u^2 - v^2, \right.$$

$$\left. 2xy, 2uv, 2xu, 2xv, 2yu, 2yv \right), \quad (x, y, u, v) \in S^3 \subset \mathbf{R}^4.$$

The story is the same as before: the components of f_{λ_2} comprise an L^2-orthonormal basis in the space of quadratic spherical harmonics $\mathcal{H}^2_{S^3}(\mathbf{R})$ and f_{λ_2} factors through the antipodal map yielding a minimal imbedding of the real projective space $\mathbf{R}P^3$ into S^8. Albeit similar, the new f_{λ_2} has much more flexibility than the old one. This is primarily due to the fact that S^3 is also the unit sphere of $\mathbf{C}^2$ so that, turning to complex, we are tempted to search for $U(2)$-invariant objects. Hence we venture to restrict the $SO(4)$-module structure on the complexification $\mathcal{H}^2_{S^3}$ to obtain the branching

$$\mathcal{H}^2_{S^3}|_{U(2)} \cong \mathcal{H}^{2,0}_{\mathbf{C}P^1} \oplus \mathcal{H}^{1,1}_{\mathbf{C}P^1} \oplus \mathcal{H}^{0,2}_{\mathbf{C}P^1},$$

where $\mathcal{H}^{p,q}_{\mathbf{C}P^1}$, $p+q = 2$, $p, q \geq 0$, stands for the (irreducible complex) $U(2)$-module of complex harmonic polynomials that are homogeneous of degree p in z, w and homogeneous of degree q in $\bar{z}, \bar{w}$. The fancy subscript of the complex projective line $\mathbf{C}P^1$ means only that any such polynomial is uniquely determined by its values taken on $\mathbf{C}P^1$.

Choosing an L^2-orthonormal basis in $\mathcal{H}_{\mathbf{C}P^1}^{2,0}$ appropriately gives rise to the complex Veronese map

$$V : S^3 \to S^5$$

defined by

$$V(z, w) = (z^2, \sqrt{2}zw, w^2), \quad (z, w) \in \mathbf{C}^2.$$

Factoring out by the diagonal center $S^1 \subset U(2)$, V projects down to what is called the holomorphic Veronese map

$$V : \mathbf{C}P^1 \to \mathbf{C}P^2$$

whose image is a quadratic curve given, in homogeneous coordinates $[X : Y : Z]$ on $\mathbf{C}P^2$, by $2XZ = Y^2$.

We now consider the middle term $\mathcal{H}_{\mathbf{C}P^1}^{1,1}$. (Note that $\mathcal{H}_{\mathbf{C}P^1}^{0,2}$ is just the dual of $\mathcal{H}_{\mathbf{C}P^1}^{2,0}$ so that it does not give new information.) An L^2-orthonormal basis in $\mathcal{H}_{\mathbf{C}P^1}^{1,1}$ gives rise to another classic: the Hopf map

$$H : S^3 \to S^2$$

defined by

$$H(z, w) = (|z|^2 - |w|^2, 2z\bar{w}), \quad z, w \in \mathbf{C}^2.$$

Note that, by birth, V and H are equivariant with respect to the $U(2)$-module structures on $\mathcal{H}_{\mathbf{C}P^1}^{2,0}$ and $\mathcal{H}_{\mathbf{C}P^1}^{1,1}$, respectively.

f_{λ_2}, V and H are some of the classical examples of eigenmaps between spheres, i.e. they share the property that their components are spherical harmonics of a fixed order. A fundamental problem in harmonic map theory posed by [Eells-Lemaire;3] is to develop a classification theory for such maps. The main objective of these notes is to study the parameter spaces called moduli spaces of eigenmaps. It will turn out that the moduli spaces, though finite dimensional, are enormous and the fine structure of their boundary is particularly subtle. Our treatment also includes the minimal immersions whose moduli spaces were first studied in the pioneering work of [DoCarmo-Wallach]. Another byproduct is a classification theory for orthogonal multiplications that are intimately related to Clifford modules.

Our framework is general harmonic map theory originated by [Eells-Sampson]. The existence of the two reports on harmonic maps by [Eells-Lemaire;1,2] (cf. also [Eells-Lemaire;3]) gives us the opportunity to be brief in the general discussion and to keep the references to the absolute minimum.

As expected, the study of the moduli spaces requires a heavy use of representation theory, especially the novel decompositions of $S^2(\mathcal{H}^{p,p}_{\mathbb{C}P^m})$ and $\mathcal{H}^{p,q}_{\mathbb{C}P^m} \otimes \mathcal{H}^{q,p}_{\mathbb{C}P^m}$ into $U(m+1)$-irreducible components. We believe that these are also of independent interest (cf. [Barbasch-Glazebrook-Toth]). Finally, it is the author's pleasure to record his thanks to Dan Barbasch for helping to develop these decomposition formulae. Without his guidance this work would never have been accomplished.

Gabor Toth

CHAPTER I

MODULI SPACES OF HARMONIC MAPS

AND MINIMAL IMMERSIONS

§1. First things first on harmonic maps and minimal immersions

Let M and N be Riemannian manifolds and $f : M \to N$ a map of class C^∞. We think of the *differential* f_* of f as a family of linear maps $f_{*x} : T_x(M) \to T_{f(x)}(N)$ parametrized by $x \in M$. We then define the *energy density* of f to be the function $e(f)$ on M which, at $x \in M$, takes the Hilbert-Schmidt norm square of f_{*x}. Concisely, we put

$$(1.1) \qquad\qquad e(f) = |f_*|^2.$$

We can also think of f_* as a 1-form on M with values in the vector bundle $f^*T(N)$ that is the pull-back of $T(N)$ via f. We can further rephrase this by saying that f_* is a section of the tensor product bundle $T^*(M) \otimes f^*T(N)$. This vector bundle carries a fibre metric that is naturally induced from the Riemannian metrics of M and N. Now (1.1) simply says that $e(f)$ is the norm square of f_* viewed as a section of $T^*(M) \otimes f^*T(N)$. Whatever way we interpret the energy density, at a point $x \in M$, $e(f)(x)$ measures the sum of the squares of stretches along an orthonormal basis in $T_x(M)$.

As the name suggests, we consider $e(f)$ as a density on M and wish to integrate it with respect to the Riemann measure ν_M of M defined by the metric on M. The

1

integral would then be called the *energy* of f and would be denoted by $E(f)$. However, for M noncompact, $e(f)$ may not be integrable. To circumvent this difficulty, we take any precompact domain D in M and define

$$(1.2) \qquad E_D(f) = \int_D e(f) \cdot \nu_M$$

to be the *energy* of f *over* D. By the Riemannian nature of the energy, it is invariant under (pre)composing f with isometries. For M a surface, it is also invariant under conformal transformations of the domain, as can easily be seen.

We are interested in those maps $f : M \to N$ whose energy is stable to first order with respect to (compactly supported) variations of f. In this case f is called a *harmonic map*. A motivation for harmonic maps is offered by Morse theory of geodesics. In fact, for $\dim M = 1$, a harmonic map $f : M \to N$ is nothing but a geodesic of N parametrized by geodesic affine parameter.

Taking an infinitesimal variation v of f, which we think of as a vector field along f or, what is the same, a section of $f^*T(N)$, we can use (1.2) to compute the derivative of the energy at the direction of v as long as we assume that the support of v lies in a precompact domain D of M. We arrive at the *first variation formula*

$$(1.3) \qquad \delta E_D(v) = -2 \int_M \langle\, \text{trace}\, \beta(f), v \rangle \cdot \nu_M\,, \ \text{supp}\, v \subset D.$$

Here $\beta(f) = \nabla f_*$, where the covariant differentiation ∇ is that of the (Riemannian connected) vector bundle $\wedge T^*(M) \otimes f^*T(N)$. In other words, $\nabla = \nabla^M \otimes f^*\nabla^N$, where ∇^M and ∇^N are the Levi-Civita covariant differentiations of M and N. $\beta(f)$ is said to be the *second fundamental form* of f; it is a symmetric 2-tensor on M with values in $f^*T(N)$. Equivalently, $\beta(f)$ is a section of $S^2 T^*(M) \otimes f^*T(N)$. We can think of the second fundamental form as the only meaningful (i.e. invariant) second derivative of the map $f : M \to N$.

By the first variation formula (1.3), $f : M \to N$ is harmonic iff

$$(1.4) \qquad \text{trace}\, \beta(f) = 0.$$

Actually, we have just written down the Euler-Lagrange equation associated to the energy. Geometrically, the *tension field* $\tau(f) = \text{trace } \beta(f)$ that appears on the left hand side of (1.4) is a vector field along f at the direction of which the energy decreases most rapidly. Analitically, τ is a quasilinear second order elliptic differential operator.

The Euler-Lagrange equation (1.4) takes a simpler form if we assume that N is the unit sphere S_V of a Euclidean vector space V (with scalar product $\langle\,,\rangle$). A map $f : M \to S_V$ can be thought of as a map $f : M \to V$ (denoted by the same symbol by abuse of terminology) with $|f|^2 = \langle f, f\rangle = 1$. We wish to think of a vector field v along f as a map $\check{v} : M \to V$ obtained form v by applying the natural shift $\check{}: T(V) \to V$. As v is tangent to S_V, we have $\langle \check{v}, f\rangle = 0$. Conversely, if $\tilde{v} : M \to V$ is a map such that $\langle \tilde{v}, f\rangle = 0$ then there exists a unique vector field v along f satisfying $\check{v} = \tilde{v}$.

For X a vector field on M, we have

$$(1.5) \qquad\qquad f_*(X)\check{} = X(f).$$

Here, on the left hand side, f maps into S_V and $T(S_V)$ is considered to be contained in $T(V)$. On the right hand side, f maps into V and the vector field X acts on f as a first order differential operator. Instead of $X(f)$ we could have written $df(X)$. Moreover, if v is a vector field along f, we have

$$(1.6) \qquad\qquad (\nabla_X v)\check{} = X(\check{v}) - \langle X(\check{v}), f\rangle \cdot f.$$

This is because $\nabla_X v$ on the left hand side is the covariant derivative of v *along* $f : M \to V$ projected down to $T(S_V)$.

PROPOSITION 1.1. *A map* $f : M \to S_V$ *is harmonic iff*

$$(1.7) \qquad\qquad \Delta^M f = e(f) \cdot f,$$

3

where $\triangle^M$ is the Laplace-Beltrami operator acting on $f : M \to V$.

PROOF: We work out the second fundamental form $\beta(f)$. Let X and Y be vector fields on M. Then, using (1.5) and (1.6), we have

$$\beta(f)(X,Y)\check{} = (\nabla_X f_*)(Y)\check{} = \nabla_X(f_*(Y))\check{} - f_*(\nabla_X Y)\check{}$$
$$= X(Y(f)) - \langle X(Y(f)), f \rangle f - (\nabla_X Y)(f)$$
$$= X(Y(f)) - (\nabla_X Y)(f) + \langle X(f), Y(f) \rangle f,$$

where the last equality is because $\langle Y(f), f \rangle = \frac{1}{2} Y(|f|^2) = 0$. Taking traces, we obtain

$$\text{trace } \beta(f)\check{} = -\triangle^M f + e(f) \cdot f$$

and (1.7) follows. $\checkmark$

Let λ be a nonzero eigenvalue of $\triangle^M$ acting on functions on M. A harmonic map $f : M \to S_V$ is said to be a λ-*eigenmap* if $e(f) = \lambda$. By Proposition 1.1, the components of f are eigenfunctions of $\triangle^M$ corresponding to the eigenvalue λ. While in this chapter we deal with harmonic maps in general our main concern will later be λ-eigenmaps.

Returning to the general situation, we wish to describe the behavior of the energy *near* a harmonic map $f : M \to N$. This is the content of the second variation formula for the energy:

$$\delta^2 E_D(v,w) = 2 \int_M \langle J_f(v), w \rangle \cdot \nu_M, \text{ supp } v \cap \text{supp } w \subset D,$$

where the *Jacobi operator* J_f is given by

$$J_f(v) = -\text{trace } \nabla^2 v + \text{trace } R^N(f_*, v) f_*.$$

Here, the positive operator ∇^2 is, for X and Y vector fields on M, defined by

$$\nabla^2_{X,Y} v = \nabla_X \nabla_Y v - \nabla_{\nabla_X Y} v.$$

4

We may think of ∇^2 as half of the curvature operator f^*R^N of $f^*T(N)$, where R^N is the Riemann curvature of N.

Analitically, J_f is a *linear* elliptic differential operator on $f^*T(N)$. If $J_f(v) = 0$, then v is said to be a *Jacobi field* along f. The *nullity* of f is the dimension of $\ker J_f$ and it measures how degenerate the energy is near f. For M compact, the nullity of f is finite by ellipticity.

Finally, the *generalized divergence* of a vector field v along f is a function on M defined by

$$\mathrm{div}_f v = \mathrm{trace}\, \langle \nabla v, f_* \rangle.$$

v is said to be *divergencefree* if $\mathrm{div}_f v = 0$.

PROBLEM: Let $f : M \to N$ be a harmonic map between Riemannian manifolds and X an infinitesimal isometry on M. Show that f_*X is a Jacobi field along f. Moreover, f_*X is divegencefree if $e(f)$ is constant.

PROPOSITION 1.2. *Let v be a vector field along a harmonic map $f : M \to S_V$, where V is a Euclidean vector space. Then, we have*

$$J_f(v)\tilde{} = \triangle^M \check{v} - e(f)\check{v} - 2\,\mathrm{div}_f v \cdot f.$$

In particular, v is a divergencefree Jacobi field along f iff

(1.8) $$\triangle^M \check{v} = e(f) \cdot \check{v}.$$

PROOF: Iterating (1.6), we obtain

$$(\nabla_X \nabla_Y v)\tilde{} = X(Y(\check{v})) - \langle X(Y(\check{v})), f \rangle f - \langle Y(\check{v}), f \rangle X(f)$$

for any vector fields X and Y on M. Taking traces

$$(\mathrm{trace}\,\nabla^2 v)\tilde{} = -\triangle^M \check{v} + \langle \triangle^M \check{v}, f \rangle f - \mathrm{trace}\, \langle d\check{v}, f \rangle df.$$

5

As S_V is a space of constant curvature 1, we have

$$(\text{trace } R^N(f_*, v)f_*)\check{} = (\text{trace }\langle f_*, v\rangle f_*)\check{} - e(f)\check{v}$$
$$= \text{trace }\langle df, \check{v}\rangle df - e(f)\check{v}$$
$$= -\text{trace }\langle f, d\check{v}\rangle df - e(f)\check{v},$$

where the last equality is because $\langle f, \check{v}\rangle = 0$. Putting these together, we obtain

$$J_f(v)\check{} = \Delta^M\check{v} - e(f)\check{v} - \langle\Delta^M\check{v}, f\rangle f.$$

It remains to show that $\langle\Delta^M\check{v}, f\rangle = 2\,\text{div}_f v$. We rewrite the left hand side as

$$\langle\Delta^M\check{v}, f\rangle = \Delta^M\langle\check{v}, f\rangle - \langle\check{v}, \Delta^M f\rangle + 2\langle d\check{v}, df\rangle.$$

Now, the first two terms on the right hand side vanish because of (1.7). On the other hand, by (1.5) and (1.6), we have $\text{div}_f v = \text{trace }\langle\nabla v, f_*\rangle = \text{trace }\langle d\check{v}, df\rangle$ which completes the proof. $\sqrt{}$

PROBLEM: Let M be a Riemannian manifold and $f : M \to S_V$ a harmonic map, where V is a Euclidean vector space.

(a) Show that the second variation of the energy can be written as

$$\delta^2 E_D(v, v) = 2\int_M (|d\check{v}|^2 - e(f)|\check{v}|^2) \cdot \nu_M,$$

where v is a vector field along f with support in $D \subset M$.

(b) Assume that M is compact. Let Z be a *uniform* vector field on V, i.e. $\check{Z} \in V$ is constant, and denote by $Z^\top$ the vector field on S_V obtained from Z by (restriction and) projection. Setting $q(\check{Z}) = \langle J_f(Z^\top \circ f), Z^\top \circ f\rangle$, derive the formula

$$q(\check{Z}) = \langle\check{Z}, f\rangle^2 e(f) + \text{trace }\langle f_*, Z\rangle^2 - |Z^\top|^2 e(f),$$

where the trace is taken with respect to the Riemannian metric on M.

6

Finally show that

$$\text{trace } q = (3 - \dim V) e(f).$$

(q is a quadratic form on V at every point of M and the trace is taken with respect to the scalar product on V.)

(A harmonic map is said to be *stable* if the second variation of the energy is positive semidefinite. The problem shows that any nonconstant harmonic map $f : M \to S_V$ is unstable provided that $\dim V \geq 4$ (cf. [Xin]).)

PROBLEM *: Let v be a vector field along a harmonic map $f : M \to S_V$. Show that v is a *harmonic variation* of f, i.e. $f_t = \exp \circ (tv) : M \to S_V$ is harmonic for all $t \in \mathbf{R}$ iff v is a divergencefree Jacobi field along f with $|v|$ constant (cf. [Toth;1]).

We now turn to minimal immersions. Let M be a (pure) manifold of dimension m and N a Riemannian manifold. Given an immersion $f : M \to N$ of class C^∞, we define $\wedge^m f_*$ to be the m-form on M with values in $\wedge^m f^* T(N)$ which, for $x \in M$, associates to the m-tuple $(X_x^1, \dots, X_x^m)$ of tangent vectors at $x \in M$ the m-vector

$$(\wedge^m f_*)(X_x^1, \dots, X_x^m) = f_*(X_x^1) \wedge \dots \wedge f_*(X_x^m) \in \wedge^m T_{f(x)}(N).$$

We define the *volume density* $v(f)$ of f to be the m-form on M that is the norm of $\wedge^m f_*$ with respect to the fibre metric of $\wedge^m f^* T(N)$ naturally induced from the Riemannian metric on N. Concisely, we put

$$v(f) = |\wedge^m f_*|.$$

For a precompact domain D in M, we define

$$V_D(f) = \int_D v(f)$$

to be the *volume of f over D*. Unlike the energy, the volume is invariant under any diffeomorphism $a : D \to D'$ of precompact domains D and D' in M, i.e. we have

(1.9)
$$V_D(f \circ a) = V_{D'}(f).$$

7

This is because $v(f \circ a) = a^* v(f)$. Formula (1.9) entails that for the description of the behavior of the volume under infinitesimal variations, we have to take into account only those which are orthogonal to the image of f. These *normal variations* are nothing but sections of the normal bundle ν_f that is given by the orthogonal splitting

(1.10) $$f^* T(N) = T(M) \oplus \nu_f.$$

We then have the first variation formula for the volume :

(1.11) $$\delta V_D(v) = - \int_M \langle \, \text{trace} \, \beta(f), v \rangle \cdot \nu_M \, , \; \text{supp} \, v \subset D,$$

where v is a normal variation of f. Here $\beta(f)$ is the second fundamental form of f when M is endowed with the Riemannian metric induced from the Riemannian metric of N by the immersion f. With respect to this metric, $f : M \to N$ is an isometric immersion. In accordance with (1.11), the second fundamental form $\beta(f)$ takes its values in the subbundle $\nu_f \subset f^* T(N)$. This allows us to define the *second fundamental form* $\beta(f)$ *of the immersion* $f : M \to N$ as a symmetric 2-tensor on M with values in ν_f, or what is the same, as a section of $S^2(T^*(M) \otimes \nu_f$.

An immersion $f : M \to N$ is said to be *minimal* if its volume is stable to first order with respect to (compactly supported) normal variations of f. The previous argument then shows that an immersion is minimal iff it is harmonic *as an isometric immersion*. For an isometric immersion $f : M \to N$, the energy density is clearly m so that we have the following:

PROPOSITION 1.3. *Let* V *be a Euclidean vector space. An immersion* $f : M \to S_V$ *is minimal iff*

(1.12) $$\triangle^M f = m \cdot f,$$

where $\triangle^M$ *is the Laplace-Beltrami operator on* M *with respect to the metric induced from* N *by* f. $\sqrt{}$

8

REMARK: An isometric minimal immersion is clearly a λ-eigenmap with $\lambda = m$.

The following result is due to [Takahashi]:

PROPOSITION 1.4. *Let V be a Euclidean vector space and $f : M \to V$ an isometric immersion such that (1.12) holds. Then the image of f is contained in S_V and the restriction $f : M \to S_V$ is a minimal immersion.*

PROOF: By Proposition 1.3, we need only to prove the first statement. Let $\beta(f)$ be the second fundamental form of f. For X and Y vector fields on M, we have

$$\beta(f)(X,Y)\check{\ } = (\nabla_X f_*)(Y)\check{\ } = \nabla_X(f_*(Y))\check{\ } - f_*(\nabla_X Y)\check{\ }$$
$$= X(Y(f)) - (\nabla_X Y)(f).$$

Taking traces, we obtain

(1.13) $$\mathrm{trace}\,\beta(f)\check{\ } = -\Delta^M f.$$

On the other hand, $\mathrm{trace}\,\beta(f)$ is a section of the normal bundle ν_f so that

$$\langle\, \mathrm{trace}\,\beta(f), f_* \rangle = \langle\, \mathrm{trace}\,\beta(f)\check{\ }, df \rangle = 0.$$

Using (1.12) and (1.13), we obtain $\langle f, df \rangle = 0$ so that $|f|^2$ is constant. Using again that $f : M \to V$ is isometric, we have

$$0 = \Delta^M |f|^2 = 2\langle\Delta^M f, f\rangle - 2\,\mathrm{trace}\,\langle df, df \rangle = 2m|f|^2 - 2m$$

and $|f|^2 = 1$ follows. $\sqrt{\ }$

For completeness, we briefly discuss the second variation formula for the volume. Let $f : M \to N$ be an (isometric) minimal immersion. Given normal fields v and w along f, we have

$$\delta^2 V_D(v, w) = \int_M \langle J_f(v), w \rangle \cdot \nu_M \,, \ \mathrm{supp}\,v \cap \mathrm{supp}\,w \subset D,$$

9

where the *normal Jacobi operator* J_f is a linear elliptic second order differential operator of ν_f given by

$$J_f(v) = -\text{trace}\nabla^2 v + \text{trace}R^N(f_*, v)f_* - \tilde{A}(f)(v).$$

Here ∇ is the covariant differentiation in ν_f induced from that of $f^*T(N)$ by projection in the splitting (1.10). Moreover, $\tilde{A}(f) = A(f)^\top \circ A(f)$ is the positive semidefinite endomorphism of ν_f, where $A(f) : \nu_f \to S^2 T(M)$ stands for the *shape operator* of f. The shape operator corresponds to the second fundamental form via the isomorphism

$$S^2 T^*(M) \otimes \nu_f \cong \text{Hom}(\nu_f, S^2 T(M)).$$

A normal vector field v along f is said to be a *normal Jacobi field along f* if $J_f(v) = 0$. $\dim \ker J_f$ is the *normal nullity* of f.

REMARK: For $N = S_V$, V a Euclidean vector space, it is possible to derive $J_f(v)\tilde{}$ in terms of $\check{v}$. We omit the details.

§2. Construction of the moduli spaces

A map $f : M \to S_V$ of class C^∞ of a Riemannian manifold M into the unit sphere S_V of a Euclidean vector space V is said to be *full* if the image of $f : M \to V$ spans V. In general, f factors through the inclusion $V_0 = \text{span}\,\{f(x)|x \in M\} \subset V$ inducing a full map $f_0 : M \to S_{V_0}$. By Proposition 1.1, if $f : M \to S_V$ is harmonic then $f_0 : M \to S_{V_0}$ is also harmonic.

Two maps $f_1 : M \to S_{V_1}$ and $f_2 : M \to S_{V_2}$ are said to be *orthogonally equivalent*, written as $f_1 \cong f_2$, if there exists a linear isometry $U : V_1 \to V_2$ between the Euclidean vector spaces V_1 and V_2 such that $U \circ f_1 = f_2$. Harmonicity and fullness are preserved by orthogonal equivalence which is then an equivalence relation on the set of all full harmonic maps $f : M \to S_V$ of a fixed Riemannian manifold into the unit sphere S_V of a Euclidean vector space V, for various V.

Let $f : M \to S_V$ and $f' : M \to S_{V'}$ be full harmonic maps. f' is said to be *derived* from f, written as $f' \leftharpoonup f$, if there exists a linear map $A : V \to V'$ such that $A \circ f = f'$. As f is full, A is uniquely determined. As f' is full, A is onto. By Proposition 1.1, if $f' \leftharpoonup f$ then $e(f') = e(f)$. In particular, if f is a λ-eigenmap then so is f'.

Our present objective is to give a parametrization of the set of orthogonal equivalence classes of full harmonic maps that are derived from a given full harmonic map $f : M \to S_V$.

To do this, let $S^2 V$ denote the Euclidean vector space of symmetric endomorphisms of V. As usual, the scalar product on $S^2 V \subset \mathrm{Hom}\,(V, V)$ is given by

$$(2.1) \qquad \langle C, C' \rangle = \mathrm{trace}\,(C'^{\mathsf{T}} \cdot C)\,,\ C, C' \in \mathrm{Hom}\,(V, V),$$

where ^{T} stands for transpose. For a nonzero vector $v \in V$, let $\mathrm{proj}\,[v] \in S^2 V$ denote the orthogonal projection of V onto $\mathbf{R} \cdot v$. For $v \in S_V$, we then have $\mathrm{proj}\,[v](w) = \langle w, v \rangle v$, $w \in V$. Since symmetric endomorphisms are diagonalizable:

$$(2.2) \qquad S^2 V = \mathrm{span}\,\{\, \mathrm{proj}\,[v] | v \in S_V \}.$$

For a given full harmonic map $f : M \to S_V$ we now define

$$(2.3) \qquad \mathcal{W}_f = \mathrm{span}\,\{\, \mathrm{proj}\,[f(x)] | x \in M \} \subset S^2 V.$$

Let $\mathcal{E}_f \subset S^2 V$ denote the orthogonal complement of $\mathcal{W}_f$ in $S^2 V$. Finally, let

$$(2.4) \qquad \mathcal{L}_f = \{ C \in \mathcal{E}_f | C + I \geq 0 \},$$

where $I = I_V =$ identity of V and '$\geq$' stands for positive semidefinite. Clearly, $\mathcal{L}_f$ is a convex body in $\mathcal{E}_f$ containing the origin in its interior.

11

THEOREM 2.1. *Given a full harmonic map* $f : M \to S_V$, *the set of orthogonal equivalence classes of full harmonic maps* $f' : M \to S_{V'}$ *that are derived from* f *can be parametrized by the convex body* $\mathcal{L}_f$. *The parametrization is given by associating to the equivalence class of* f' *the symmetric endomorphism*

$$(2.5) \qquad\qquad \langle f' \rangle_f = A^\top \cdot A - I$$

of V, *where* $f' = A \circ f$.

PROOF: Clearly, $\langle f' \rangle_f \in S^2 V$ depends only on the equivalence class of f'. By (2.1), for $x \in M$, we have

$$
\begin{aligned}
\langle \langle f' \rangle_f, \, \text{proj}\,[f(x)] \rangle &= \text{trace}\,(\,\text{proj}\,[f(x)]\langle f' \rangle_f) \\
&= \langle (A^\top A - I)f(x), f(x) \rangle \\
&= |Af(x)|^2 - |f(x)|^2 \\
&= |f'(x)|^2 - 1 = 0
\end{aligned}
$$

so that $\langle f' \rangle_f \in \mathcal{E}_f$. As $A^\top A \geq 0$ for any linear map A, we actually have $\langle f' \rangle_f \in \mathcal{L}_f$. To show that the parametrization is injective, let $f_l : M \to S_{V_l}$, $l = 1, 2$, be derived from f via $f_l = A_l \circ f$, $A_l : V \to V_l$, and assume that $\langle f_1 \rangle_f = \langle f_2 \rangle_f$. This equality translates into

$$(2.6) \qquad\qquad A_1^\top A_1 = A_2^\top A_2,$$

in particular, the linear maps A_1 and A_2 have the same kernel $K \subset V$. We may assume that K is zero (or else, restrict A_1 and A_2 to $K^\perp \subset V$). By polar decomposition, for $l = 1, 2$, we have $A_l = U_l \cdot Q_l$, where $U_l : V \to V_l$ is a linear isometry and Q_l is a symmetric positive definite endomorphism of V. Substituting this into (2.6) and taking the square root of both sides, we obtain $Q_1 = Q_2$. Now, we have $f_1 = A_1 \circ f = U_1 Q_1 \circ f = U_1 Q_2 \circ f = U_1 U_2^{-1} \circ f_2$ so that f_1 and f_2 are orthogonally equivalent.

12

To show that the parametrization is surjective, given $C \in \mathcal{L}_f$, define $A = \sqrt{C + I}$: $V \to V$. Let $A_0 : V \to V_0$, $V_0 = $ image of A, denote the linear map obtained from A by restriction. Then $f_0 = A_0 \circ f : M \to S_{V_0}$ is a full harmonic map derived from f with $\langle f_0 \rangle_f = A_0^\top A_0 - I = A^2 - I = C$. $\checkmark$

The convex body $\mathcal{L}_f$ of $\mathcal{E}_f (\subset S^2 V)$ is said to be the *moduli space* associated to the full harmonic map $f : M \to S_V$. By (2.5), f corresponds to the origin $0 (= \langle f \rangle_f)$. It also follows that the interior $\mathcal{L}_f^\circ$ parametrizes those full harmonic maps $f' : M \to S_{V'}$ for which $f \leftharpoondown f'$, or equivalently, $f' = A \circ f$ with $A : V \to V'$ invertible. As A is onto, this holds iff $\dim V = \dim V'$.

EXAMPLE 2.2: Let $f : M \to S_V$ be a harmonic map and assume that the image of f has nonempty interior in S_V. Then $\mathcal{L}_f = \mathcal{E}_f = \{0\}$. In fact, let $f' \leftharpoondown f$ via $f' = A \circ f$, where $A : V \to V'$ is a surjective linear map. By assumption, A preserves the norm on an open set of S_V so that it is a linear isometry.

THEOREM 2.3. *If M is compact then $\mathcal{L}_f$ is compact for any full harmonic map $f : M \to S_V$.*

PROOF: Integrating $\mathrm{proj}\,[f] : M \to S^2 V$ over M, let

$$Q = \int_M \mathrm{proj}\,[f] \cdot \nu_M \in S^2 V,$$

where ν_M is the Riemann measure on M. We first claim that Q is positive definite. Indeed, for $v \in V$, we have

$$\langle Qv, v \rangle = \int_M \langle \mathrm{proj}\,[f] v, v \rangle \cdot \nu_M = \int_M \langle f, v \rangle^2 \cdot \nu_M \geq 0$$

and equality holds iff $\langle f, v \rangle = 0$, i.e. iff the image of f is contained in $v^\perp$. As f is full, we obtain $v = 0$.

Secondly, we claim that, for $C \in \mathcal{L}_f$, we have

(2.7) $$\mathrm{trace}\, C \leq \frac{\mathrm{vol}\,(M)}{\lambda_{\min}} - \dim V,$$

13

where $\mathrm{vol}\,(M) = \int_M \nu_M$ is the volume of M and $\lambda_{\min} > 0$ is the smallest eigenvalue of Q. Indeed, for $C \in \mathcal{E}_f$, integrating the defining equality $\langle C, \mathrm{proj}\,[f] \rangle = 0$ over M we obtain $\langle C, Q \rangle = 0$. If, in addition, $C + I \geq 0$, then we have

$$\mathrm{vol}\,(M) = \int_M |f|^2 \cdot \nu_M = \int_M \langle I, \mathrm{proj}\,[f] \rangle \cdot \nu_M = \langle I, Q \rangle = \langle C + I, Q \rangle$$
$$= \mathrm{trace}\,(Q \cdot (C + I)) \geq \lambda_{\min}\,(\,\mathrm{trace}\,(C + I)) = \lambda_{\min}\,(\,\mathrm{trace}\,C + \dim V),$$

where the inequality can be obtained by expressing the trace in terms of an orthonormal basis consisting of eigenvectors of Q. Thus estimate (2.7) follows.

On the other hand, since $C + I \geq 0$, the eigenvalues of C are ≥ -1. Combining this with (2.7) we obtain that the eigenvalues of the elements in $\mathcal{L}_f$ are bounded. Thus $\mathcal{L}_f$ is compact. $\checkmark$

REMARK: If

$$(2.8) \qquad \int_M \mathrm{proj}\,[f] \cdot \nu_M = \lambda \cdot I$$

then the proof above shows that, for any $C \in \mathcal{L}_f$, we have

$$(2.9) \qquad \mathrm{trace}\,C = \frac{\mathrm{vol}\,(M)}{\lambda} - \dim V.$$

Let $f : M \to S_V$ be a full harmonic map and assume that $f' : M \to S_{V'}$ is derived from f by the linear map $A : V \to V'$, i.e. $f' = A \circ f$. We introduce the affine map

$$\iota : S^2 V' \to S^2 V$$

by

$$(2.10) \qquad \iota(C') = A^\top \cdot C' \cdot A + \langle f' \rangle_f = A^\top \cdot (C' + I_{V'}) \cdot A - I_V\,,\ C' \in S^2 V'.$$

PROPOSITION 2.4. *The map ι is injective and maps $\mathcal{E}_{f'}$ into $\mathcal{E}_f$. Moreover, we have*

$$(2.11) \qquad \iota(\mathcal{L}_{f'}) = \iota(\mathcal{E}_{f'}) \cap \mathcal{L}_f.$$

PROOF: Let $C_1, C_2 \in S^2 V'$ be such that $\iota(C_1) = \iota(C_2)$, or equivalently, $A^\top(C_1 - C_2)A = 0$. Since $A : V \to V'$ is onto, $A^\top : V' \to V$ is injective so that $C_1 - C_2 = 0$ follows.

Given $C' \in \mathcal{E}_{f'}$, we have

$$\langle \iota(C'), \text{proj}\,[f] \rangle = \langle \iota(C')f, f \rangle = \langle (C' + I_{V'})A \circ f, A \circ f \rangle - 1$$

$$= \langle C' \circ f', f' \rangle = \langle C', \text{proj}\,[f'] \rangle = 0$$

and hence $\iota(\mathcal{E}_{f'}) \subset \mathcal{E}_f$.

Finally, for $C' \in \mathcal{E}_{f'}$, we have $C' \in \mathcal{L}_{f'}$ iff $C' + I_{V'} \geq 0$. Since $A : V \to V'$ is onto, this is equivalent to

$$\iota(C') + I_V = A^\top(C' + I_{V'})A \geq 0$$

and (2.11) follows.$\sqrt{}$

For $f' \leftharpoonup f$, we define

$$I_{f'} = \iota(\mathcal{L}_{f'}^\circ) \subset \mathcal{L}_f.$$

Clearly, $I_{f'}$ is convex and open in $\iota(\mathcal{E}_{f'})$. Moreover, $\langle f' \rangle_f = \iota(0) \in I_{f'}$. For $f = f'$, the affine map ι is the identity so that we have $I_f = \mathcal{L}_f^\circ$. $I_{f'}$ is said to be the *cell associated to* f'.

PROPOSITION 2.5. *The closure* $\bar{I}_{f'} = \iota(\mathcal{L}_{f'}) \subset \mathcal{L}_f$ *parametrizes those full harmonic maps* f'' *for which* $f'' \leftharpoonup f'(\leftharpoonup f)$.

PROOF: Given $\iota(C') \in \bar{I}_{f'}$ with $C' \in \mathcal{L}_{f'}$, we define

$$f'' = \sqrt{C' + I_{V'}} \circ f'$$

(and restrict the range of f'' to a full harmonic map denoted by the same symbol). Then $f'' \leftharpoonup f'$ and we have

$$\langle f'' \rangle_f = A^\top(C' + I_{V'})A - I_V = \iota(C').$$

15

The converse is similar. $\checkmark$

The proof above shows that the cell $I_{f'}$ associated to f' parametrizes those full harmonic maps f'' for which $f'' \rightleftharpoons f'$, i.e. f' and f'' are derived from each other. In particular, for $\langle f'' \rangle_f \in \partial I_{f'} = \bar{I}_{f'} \setminus I_{f'}$, the range dimension of f'' is strictly less than that of f'.

Since the symmetrized relation $\rightleftharpoons$ is an equivalence, $\mathcal{I}_f = \{I_{f'} | f' \leftharpoondown f\}$ is a decomposition of the moduli space $\mathcal{L}_f$ into disjoint convex sets. We call $\mathcal{I}_f$ the *natural saturation* of $\mathcal{L}_f$. By construction, for $f' \leftharpoondown f$, the restriction $\iota | \mathcal{L}_{f'} : \mathcal{L}_{f'} \to \mathcal{L}_f$ preserves the natural saturations on $\mathcal{L}_{f'}$ and $\mathcal{L}_f$.

Let $f : M \to S_V$ be a full harmonic map. In what follows, we give a geometric interpretation of $\mathcal{E}_f$.

Let v be a vector field along f. Then v is said to be *derived* from f is there exists a linear endomorphism $A : V \to V$ such that $A \circ f = \check{v}$, where $\check{v} : M \to V$ is obtained by composing v with the natural shift $\check{} : T(V) \to V$. (Note that A may not be onto since the image of $\check{v}$ may not span V.) By Proposition 1.2, if v is derived from f then v is a divergencefree Jacobi field along f. This follows by applying A to both sides of (1.7) to obtain (1.8).

Writing $A = B + C$ with $B \in so(V)$ and $C \in S^2 V$, we have $C \in \mathcal{E}_f$. This is because $\langle A \circ f, f \rangle = \langle \check{v}, f \rangle = 0$ so that, using the skew-symmetry of B, we obtain $\langle C, \text{proj} [f] \rangle = \langle C \circ f, f \rangle = 0$.

PROPOSITION 2.6. *The linear space of all vector fields along f that are derived from f is contained in the space of all divergencefree Jacobi fields along f and is isomorphic with the direct sum*

$$so(V) \oplus \mathcal{E}_f.$$

In particular, we have the lower estimate

(2.12) $$\dim \ker J_f \geq \dim \mathcal{E}_f + \frac{1}{2} n(n+1)$$

16

for the nullity of f, where we put $\dim V = n + 1$. $\sqrt{}$

REMARK: Let v be derived from $f : M \to S_V$ via $A \circ f = \check{v}$. Decompose $A = B + C$ with $B \in so(V)$ and $C \in \mathcal{E}_f$ and choose $\varepsilon > 0$ such that $2tC + I > 0$ for $|t| < \varepsilon$. Define

$$f_t = e^{tB} \cdot \sqrt{2tC + I} \circ f, \ |t| < \varepsilon.$$

Then $f_t : M \to S_V$ is a full harmonic map that is derived from f. Indeed, we have

$$|f_t|^2 = |\sqrt{2tC + I} \circ f|^2 = \langle (2tC + I) \circ f, f \rangle = 2t \langle C, \text{proj}\,[f] \rangle + |f|^2 = 1.$$

Moreover, differentiating $in\ V$, we have

$$\frac{\partial f_t}{\partial t}\Big|_{t=0} = B \circ f + C \circ f = \check{v}.$$

Summarizing, we obtained that for those v that are derived from $f : M \to S_V$ there exists a 1-parameter family $f_t : M \to S_V$ of full harmonic maps such that $\partial f_t / \partial t|_{t=0} = v$. (In particular, it again follows that v is a Jacobi field along f.) This fails to hold in general for any Jacobi field along a harmonic map $f : M \to N$ between Riemannian manifolds (cf. [Smith]).

Now let $f : M \to S_V$ be a full minimal immersion. Keeping the Riemannian metric on M induced by f fixed, we give a parametrization of orthogonal equivalence classes of full $isometric$ minimal immersions $f' : M \to S_{V'}$ that are derived from f. To do this, we define

(2.13) $\qquad \mathcal{Z}_f = \text{span}\, \{\, \text{proj}\,[f_*(X_x)^\cdot] | X_x \in T_x(M)\,, \ x \in M \} \subset S^2 V.$

Let $\mathcal{F}_f \subset S^2 V$ be the orthogonal complement of $\mathcal{Z}_f$ in $S^2 V$.
Finally, put

$$\mathcal{M}_f = \{C \in \mathcal{F}_f | C + I \geq 0\}.$$

THEOREM 2.7. We have

(2.14) $\qquad\qquad\qquad\qquad \mathcal{F}_f \subset \mathcal{E}_f.$

17

PROOF: Let $C \in \mathcal{F}_f$ and choose $\varepsilon > 0$ such that $\varepsilon C + I > 0$ and define $g = \sqrt{\varepsilon C + I} \circ f : M \to V$. Applying the linear map $\sqrt{\varepsilon C + I}$ to both sides of (1.12), we obtain

$$\triangle^M g = m \cdot g.$$

We claim that g is an isometric immersion. Indeed, for $X_x \in T_x(M)$, $x \in M$, we have

$$
\begin{aligned}
|g_*(X_x)|^2 = |g_*(X_x)^\backprime|^2 &= |\sqrt{\varepsilon C + I} \cdot f_*(X_x)^\backprime|^2 \\
&= \langle (\varepsilon C + I) f_*(X_x)^\backprime, f_*(X_x)^\backprime \rangle \\
&= \varepsilon \langle C, \operatorname{proj}[f_*(X_x)^\backprime] \rangle + |f_*(X_x)^\backprime|^2 \\
&= |f_*(X_x)|^2 = |X_x|^2.
\end{aligned}
$$

We are now in the position to apply Proposition 1.4 to conclude that the image of g is contained in S_V. Hence

$$1 = |g|^2 = |\sqrt{\varepsilon C + I} \circ f|^2 = \langle (\varepsilon C + I) \circ f, f \rangle = \varepsilon \langle C, \operatorname{proj}[f] \rangle + 1$$

and $\langle C, \operatorname{proj}[f] \rangle = 0$ follows. This means that $C \in \mathcal{E}_f$ which was to be proven. $\sqrt{}$

Comparing (2.4) and (2.14), by Theorem 2.7, we get

(2.15) $$\mathcal{M}_f = \mathcal{F}_f \cap \mathcal{L}_f,$$

in particular, $\mathcal{M}_f$ is a convex body in $\mathcal{F}_f$ which, by Theorem 2.3, is compact for M compact.

THEOREM 2.8. *Given a full isometric minimal immersion $f : M \to S_V$, the set of orthogonal equivalence classes of full isometric minimal immersions $f' : M \to S_{V'}$ that are derived from f can be parametrized by the convex body $\mathcal{M}_f$.*

PROOF: Since an isometric immersion $f' : M \to S_{V'}$ is minimal iff it is harmonic, we have only to show that $\mathcal{M}_f$ parametrizes those full harmonic maps $f' : M \to S_{V'}$

18

which are isometric with respect to the metric on M induced by f. Let $f' \leftharpoondown f$ via $f' = A \circ f$. Then, we have

$$\langle \langle f' \rangle_f, \operatorname{proj}[f_*(X_x)\check{\ }] \rangle = \operatorname{trace}(\operatorname{proj}[f_*(X_x)\check{\ }]\langle f' \rangle_f)$$
$$= \langle (A^\top A - I) \cdot f_*(X_x)\check{\ }, f_*(X_x)\check{\ } \rangle$$
$$= |A \cdot f_*(X_x)\check{\ }|^2 - |f_*(X_x)|^2$$
$$= |f'_*(X_x)|^2 - |X_x|^2$$

which is zero for all $X_x \in T_x(M)$, $x \in M$, iff $f' : M \to S_{V'}$ is isometric iff $\langle f' \rangle_f \in \mathcal{F}_f$. $\checkmark$

The convex body $\mathcal{M}_f$ is said to be the *moduli space* associated to the full minimal immersion $f : M \to S_V$. By (2.15) $\mathcal{M}_f$ carries a *natural saturation* induced from that of $\mathcal{L}_f$. As in the case of harmonic maps, $\mathcal{F}_f$ can be viewed as a linear subspace of normal Jacobi fields along f. Indeed, for $C \in \mathcal{M}_f$, we have $\langle C \circ f, f \rangle = \langle C, \operatorname{proj}[f] \rangle = 0$ so that we can define the vector field v along f by putting $\check{v} = C \circ f$. Moreover, v is normal since

$$\langle v, f_* \rangle = \langle \check{v}, df \rangle = \langle C \circ f, df \rangle = \frac{1}{2} d \langle C \circ f, f \rangle = 0.$$

Finally, let $\varepsilon > 0$ be such that $2tC + I > 0$, for $|t| < \varepsilon$, and define

$$f_t = \sqrt{2tC + I} \circ f : M \to V.$$

Then, for $|t| < \varepsilon < 1/2$, $\langle f_t \rangle = 2tC \in \mathcal{M}_f$, and so $f_t : M \to S_V$ is an isometric minimal immersion. Differentiating, we obtain $\partial f_t / \partial t|_{t=0} = v$ and so v is a normal Jacobi field along f as it arises from a variation of f through isometric minimal immersions.

§3. Symmetries of the moduli space

Let $f : M \to S_V$ be a full harmonic map of a Riemannian manifold into the unit sphere S_V of a Euclidean vector space V. Assume that f is *equivariant* with respect to a homomorphism $\rho_f : G \to O(V)$, where G is a closed subgroup of the group of isometries of M. Equivariance means that, for $a \in G$, we have

$$(3.1) \qquad\qquad f \circ a = \rho_f(a) \circ f.$$

Since f is full, ρ_f is uniquely determined.

REMARK: If G acts transitively on M then the energy density $e(f)$ of f is constant so that f is a λ-eigenmap with $\lambda = e(f)$.

Returning to the general situation, the homomorphism ρ_f defines a G-module structure on V that is *orthogonal*, i.e. it preserves the scalar product $\langle\,,\,\rangle$ on V. On the symmetric square $S^2 V$ we take the induced (orthogonal) G-module structure, i.e. for a symmetric endomorphism $C : V \to V$ we define

$$a \cdot C = Ad(\rho_f(a)) \cdot C = \rho_f(a) \cdot C \cdot \rho_f(a)^\mathsf{T}, \, a \in G.$$

PROPOSITION 3.1. *Let $f : M \to S_V$ be a full harmonic map and assume that f is equivariant with respect to a homomorphism $\rho_f : G \to O(V)$. Then $\mathcal{E}_f$ is a G-submodule of $S^2 V$. Moreover, the moduli space $\mathcal{L}_f$ associated to f is a G-invariant subspace and, for $f' : M \to S_{V'}$, $f' \leftharpoondown f$, we have*

$$(3.2) \qquad\qquad a \cdot \langle f' \rangle_f = \langle f' \circ a^{-1} \rangle_f, \, a \in G.$$

PROOF: We first claim that, for $v \in S_V$, we have

$$(3.3) \qquad\qquad a \cdot \mathrm{proj}\,[v] = \mathrm{proj}\,[\rho_f(a)v], \, a \in G.$$

Indeed, for $w \in V$, we compute

$$\begin{aligned}
(a \cdot \mathrm{proj}\,[v])w &= \rho_f(a)(\,\mathrm{proj}\,[v](\rho_f(a)^\mathsf{T} w)) \\
&= \rho_f(a)(\langle \rho_f(a)^\mathsf{T} w, v \rangle v) \\
&= \langle w, \rho_f(a)v \rangle \rho_f(a)v \\
&= \mathrm{proj}\,[\rho_f(a)v]w
\end{aligned}$$

20

and (3.3) follows. By (2.3), $\mathcal{W}_f$ is a G-submodule of S^2V and so is its orthogonal complement $\mathcal{E}_f$. This proves the first statement. Furthermore, $Ad\,\rho_f$ preserves positive semidefiniteness and we infer that $\mathcal{L}_f \subset \mathcal{E}_f$ is invariant under the action of G on $\mathcal{E}_f$. Actually, for $f' \leftharpoonup f$ with $f' = A \circ f$, we have

$$a \cdot \langle f' \rangle_f = a \cdot (A^\mathsf{T}A - I) = \rho_f(a)A^\mathsf{T}A\rho_f(a)^\mathsf{T} - I$$
$$= (A\rho_f(a)^\mathsf{T})^\mathsf{T} \cdot (A\rho_f(a)^\mathsf{T}) - I = \langle f' \circ a^{-1} \rangle_f,$$

since $f' \circ a^{-1} = A \circ f \circ a^{-1} = A\rho_f(a)^\mathsf{T} \circ f.$ $\checkmark$

By (3.2), the isotropy subgroup $G_{\langle f' \rangle_f}$ of G at $\langle f' \rangle_f$ coincides with the *symmetry group*

$$G_{f'} = \{a \in G | \text{ there exists } U \in O(V') \text{ such that } f' \circ a = U \circ f'\}$$

of f'.

If $f' : M \to S_{V'}$ and $f'' : M \to S_{V''}$ are full harmonic maps with $f'' \leftharpoonup f'$ then by the very definition of derivation, for $a \in G$, we also have $f'' \circ a^{-1} \leftharpoonup f' \circ a^{-1}$. Since the natural saturation $\mathcal{I}_f$ of $\mathcal{L}_f$ is defined in terms of $\leftharpoonup$, it follows that the action of G on $\mathcal{L}_f$ respects $\mathcal{I}_f$. More precisely, by (3.2), for $f' \leftharpoonup f$ we have

$$a \cdot \mathcal{I}_{f'} = \mathcal{I}_{f' \circ a^{-1}}, \, a \in G.$$

We can thus consider G acting on $\mathcal{I}_f$.

Two full harmonic maps $f' : M \to S_{V'}$ and $f'' : M \to S_{V''}$ with $f', f'' \leftharpoonup f$ are said to be *geometrically distinct* if, for each $a', a'' \in G$, $U' \in O(V')$ and $U'' \in O(V'')$, none of the full harmonic maps

$$U' \circ f' \circ a' \quad \text{and} \quad U'' \circ f'' \circ a''$$

can be derived from the other. In terms of the natural saturation $\mathcal{I}_f$ on $\mathcal{L}_f$, this holds iff none of the orbits

$$G \cdot \bar{\mathcal{I}}_{f'} \quad \text{and} \quad G \cdot \bar{\mathcal{I}}_{f''}$$

21

is contained in the other.

Assume now that the orbit space $\mathcal{I}_f/G$ is infinite with cardinality Σ. This means that there exists an infinite set $f_\sigma : M \to S_{V_\sigma}$, $\sigma \in \Sigma$, of full harmonic maps such that the orbits $G \cdot I_{f_\sigma}$ are disjoint. Since $\dim V_\sigma \leq \dim V$, passing to a subset of the same cardinality and using the same notation, we may assume that $\dim V_\sigma$ does not depend on $\sigma \in \Sigma$. This means that the full harmonic maps f_σ are pairwise geometrically distinct. Summarizing, we obtained that if $\mathcal{I}_f/G$ is infinite with cardinality Σ then there exist Σ geometrically distinct full harmonic maps derived from f.

A full harmonic map $f : M \to S_V$ is said to be *(linearly) rigid* if $f' \leftharpoondown f$ implies $f' \cong f$, or equivalently, $\mathcal{L}_f = \mathcal{E}_f = \{0\}$.

Given $f : M \to S_V$ the rigid full harmonic maps $f' : M \to S_{V'}$ that are derived from f are exactly those for which $\bar{I}_{f'} = \{\langle f' \rangle\}$ is a 1-point cell. If $f : M \to S_V$ is equivariant with respect to $\rho_f : G \to O(V)$ and $f', f'' \leftharpoondown f$ are rigid then f' and f'' are geometrically distinct iff

$$U' \circ f' \circ a' \neq f''$$

for all $U' \in O(V')$ and $a' \in G$.

THEOREM 3.2. *Let $f : M \to S_V$ be a full harmonic map that is equivariant with respect to a homomorphism $\rho_f : G \to O(V)$. Assume that*

$$(3.4) \qquad\qquad \dim \mathcal{L}_f > \dim V(\dim G + 1).$$

Then the cardinality of $\mathcal{I}_f/G$ is $\aleph_1$. In particular, there exist $\aleph_1$ geometrically distinct full harmonic maps that are derived from f. Actually, there exist $\aleph_1$ geometrically distinct rigid full harmonic maps that are derived from f.

PROOF: Assume that $\mathcal{I}_f/G$ is countable, i.e. $\leq \aleph_0$. Then the set of G-orbits of cells on the boundary $\partial \mathcal{L}_f$ is also countable. By the Baire Category Theorem, at least

22

one G-orbit of a cell has nonempty interior in $\partial \mathcal{L}_f$. Let $\langle f_1 \rangle_f$ be an interior point. Then $f_1 : M \to S_{V_1}$ is a full harmonic map with $\dim V_1 \leq \dim V - 1$. Moreover, we have

$$(3.5) \qquad \dim \mathcal{L}_f = \dim (G \cdot I_{f_1}) + 1 \leq \dim G + 1 + \dim I_{f_1}.$$

We now take the set of G-orbits of cells which lie on the boundary $\partial I_{f_1} = \bar{I}_{f_1} \setminus I_{f_1}$. The intersections of these G-orbits with ∂I_{f_1} give a partition of ∂I_{f_1} into countably many subsets. Again by the Baire Category Theorem at least one G-orbit of a cell intersects ∂I_{f_1} in a set with nonempty interior in ∂I_{f_1}. Let $\langle f_2 \rangle$ be an interior point. Then $f_2 : M \to S_{V_2}$ is a full harmonic map with $\dim V_2 \leq \dim V_1 - 1$. Furthermore, I_{f_2} is contained in this intersection so that we have

$$\dim I_{f_1} = \dim (G \cdot I_{f_2} \cap I_{f_1}) + 1 \leq \dim (G \cdot I_{f_2}) + 1 \leq \dim G + 1 + \dim I_{f_2}.$$

Combining this with (3.5), we obtain

$$\dim \mathcal{L}_f \leq 2(\dim G + 1) + \dim I_{f_2}$$

and

$$\dim V_2 \leq \dim V - 2.$$

Repeating this, in the n-th step we obtain

$$\dim \mathcal{L}_f \leq n(\dim G + 1) + \dim I_{f_n}$$

and

$$\dim V_n \leq \dim V - n.$$

The procedure clearly stops in $\dim V$ steps yielding

$$\dim \mathcal{L}_f \leq \dim V(\dim G + 1)$$

which contradicts to (3.4).

To prove the second statement, choose geometrically distinct full harmonic maps $f_\sigma : M \to S_{V_\sigma}$, $\sigma \in \Sigma$, with Σ of cardinality $\aleph_1$. For $\sigma \in \Sigma$, choose a finite set $\Lambda_\sigma \subset \bar{I}_{f_\sigma} (\subset \mathcal{L}_f)$ consisting of points that correspond to rigid full harmonic maps such that the affine span of Λ_σ is equal to that of $\bar{I}_{f_\sigma}$. The existence of Λ_σ follows easily by induction with respect to the dimension of cells in the natural saturation of $\bar{I}_{f_\sigma}$.

We now claim that for σ, $\sigma' \in \Sigma$, $\sigma \neq \sigma'$, we have $\Lambda_\sigma \neq \Lambda_{\sigma'}$. In fact, $\Lambda_\sigma = \Lambda_{\sigma'}$ iff span $I_{f_\sigma} = $ span $I_{f_{\sigma'}}$, iff $I_{f_\sigma} = I_{f_{\sigma'}}$, so that f_σ and $f_{\sigma'}$ are not geometrically distinct. We obtain that the set $\{\Lambda_\sigma | \sigma \in \Sigma\}$ has cardinality $\aleph_1$. As the set of all finite subsets of a countable set is countable, $\cup_{\sigma \in \Sigma} \Lambda_\sigma$ has cardinality $\aleph_1$ and we are done. $\surd$

REMARK: As a byproduct of the forthcoming main result, we will obtain that (3.4) holds for infinitely many standard minimal immersions between spheres.

We now show an important property of the action of G on the natural saturation $\mathcal{I}_f$.

PROPOSITION 3.3. *Let $f : M \to S_V$ be a full harmonic map derived from f and $a : \mathbf{R} \to G$ a 1-parameter subgroup. If the orbit $s \to a(s) \cdot \langle f' \rangle_f$, $s \in \mathbf{R}$, is tangent to $I_{f'}$ at $s = 0$ then it is contained in $I_{f'}$.*

PROOF: For $s \in \mathbf{R}$, we put

$$U(s) = \rho_f(a(s)) = e^{sB} \in O(V),$$

where $B = d/dsU(s)|_{s=0} \in so(V)$. Setting $f' = A \circ f$, where $A : V \to V'$ is a linear map, by (3.2), we have

$$a(s) \cdot \langle f' \rangle_f = \langle f' \circ a(s)^{-1} \rangle_f = U(s)A^\top AU(s)^\top - I.$$

Differentiating this with respect to s at zero, we obtain

$$(3.6) \qquad \frac{d}{ds}(a(s) \cdot \langle f' \rangle_f)_{s=0} = BA^\top A - A^\top AB.$$

24

The hypothesis along with (2.10) implies that, for some $C' \in \mathcal{E}_{f'}$, we have

$$(3.7) \qquad\qquad BA^{\mathsf{T}}A - A^{\mathsf{T}}AB = A^{\mathsf{T}}C'A.$$

Linear algebra then entails that there exists a linear endomorphism $B' : V' \to V'$ satisfying

$$(3.8) \qquad\qquad AB = B'A$$

Indeed, (3.7) implies that B leaves $K = \ker A$ invariant since A^{T} is injective. Since B is skew it also leaves $K^{\perp}$ invariant. On the other hand, $A|K^{\perp} : K^{\perp} \to V'$ is a linear isomorphism since A is onto. We then define

$$B' = (AB)|K \cdot (A|K^{\perp})^{-1}$$

which clearly satisfies (3.8).

Exponentiating, (3.8) gives

$$Ae^{-sB} = AU(s)^{\mathsf{T}} = e^{-sB'}A.$$

Precomposing this with f, we obtain

$$f' \circ a(s)^{-1} = e^{-sB'} \circ f',$$

in particular, $f' \circ a(s)^{-1} \leftharpoonup f'$. By Proposition 2.5, we have $a(s) \cdot \langle f' \rangle_f = \langle f' \circ a(s)^{-1} \rangle_f \in I_{f'}$ and the proof is complete. $\checkmark$

PROBLEM: Choose $\langle f' \rangle_f$ to be the center of mass of $I_{f'}$. Use Proposition 3.3 to show that the tangent spaces of $I_{f'}$ and $G \cdot \langle f' \rangle_f$ at $\langle f' \rangle_f$ intersect trivially.

Now let $f : M \to S_V$ be a full isometric minimal immersion and assume that f is equivariant with respect to a homomorphism $\rho_f : G \to O(V)$. By (3.1) and (3.3), for $X_x \in T_x(M)$, $x \in M$, and $a \in G$, we have

$$a \cdot \mathrm{proj}\,[f_*(X_x)^{\check{}}] = \mathrm{proj}\,[\rho_f(a)f_*(X_x)^{\check{}}]$$
$$= \mathrm{proj}\,[(f \circ a)_*(X_x)^{\check{}}]$$
$$= \mathrm{proj}\,[f_*(a_*X_x)^{\check{}}].$$

25

We obtain that $\mathcal{Z}_f$ (given in (2.13)) and hence its orthogonal complement $\mathcal{F}_f$ are G-submodules of $S^2 V$ and the convex body $\mathcal{M}_f \subset \mathcal{F}_f$ is G-invariant. As to the latter, the action of G on $\mathcal{M}_f$ is, for $f' \leftharpoondown f$ and $a \in G$, given by (3.2). It follows that all the previous constructions for full harmonic maps apply to full isometric minimal immersions.

§4. The Hopf map and other examples

It is time to give some nontrivial examples. Let $\mathbf{H} = \mathbf{R}^4$ denote the skew field of *quaternions* [Berger]. With respect to the standard basis $\{1, i, j, k\}$, we have

$$q = a \cdot 1 + b \cdot i + c \cdot j + d \cdot k \in \mathbf{H}, \; a, b, c, d \in \mathbf{R}.$$

The *conjugate* of q can then be written as

$$\bar{q} = a \cdot 1 - b \cdot i - c \cdot j - d \cdot k$$

so that

$$q\bar{q} = |q|^2 = a^2 + b^2 + c^2 + d^2.$$

With respect to the quaternionic multiplication, the *symplectic group* $Sp(1)$ is just the unit sphere $S^3 \subset \mathbf{H}$. This group also acts on the (real) linear subspace

$$\mathbf{H}_0 = \{q \in \mathbf{H} | \bar{q} = -q\} = \text{span}\,\{i, j, k\} = \mathbf{R}^3$$

of *purely imaginary quaternions* by inner automorphisms. This action actually gives rise to an epimorphism $\pi : Sp(1) \rightarrow SO(3)$ with kernel $\{\pm 1\}$. By definition, for $q \in Sp(1)$, we have

$$\pi(q) \cdot q_0 = q \cdot q_0 \cdot q^{-1} = q \cdot q_0 \cdot \bar{q}, \; q_0 \in \mathbf{H}_0.$$

We define the *Hopf map*

$$H : S^3 \to S^2$$

by

(4.1)
$$H(q) = \pi(q) \cdot i = q \cdot i \cdot \bar{q}, \; q \in S^3.$$

Clearly, H is equivariant with respect to π, where $Sp(1)$ is considered to act on itself, i.e. on S^3, by left quaternionic multiplication.

$\mathbf{C} = \mathrm{span}_{\mathbf{R}}\{1, i\}$ is a complex subfield of $\mathbf{H}$ and we can write

(4.2)
$$q = z + k \cdot w, \; z, w \in \mathbf{C}.$$

As complex vector spaces, $\mathbf{H} = \mathbf{C}^2$, and left multiplication by q given in (4.2) on $\mathbf{H}$ corresponds to multiplication by the special orthogonal matrix

$$\begin{bmatrix} \alpha & -\bar{\beta} \\ \beta & \bar{\alpha} \end{bmatrix}.$$

We have just established the isomorphism $Sp(1) \cong SU(2)$.

We now work out the Hopf map $H : S^3 \to S^2$ in complex coordinates $(z, w) \in S^3 \subset \mathbf{C}^2$, $|z|^2 + |w|^2 = 1$. By (4.1) and (4.2), we have

$$H(z, w) = (z + kw)i(\bar{z} - kw)$$
$$= (|z|^2 - |w|^2)i + 2z\bar{w}j.$$

We obtain that

(4.3)
$$H(z, w) = (|z|^2 - |w|^2, 2z\bar{w}), \; |z|^2 + |w|^2 = 1.$$

The components of H are invariant under the action of the center

(4.4)
$$S^1 = \{ \mathrm{diag}\,(e^{i\theta}, e^{i\theta}) \,|\, \theta \in \mathbf{R}\}.$$

27

of the unitary group $U(2)$ so that the Hopf map is actually equivariant with respect to the homomorphism $\rho_H : U(2) \to O(3)$ that is the natural projection $U(2) \to U(2)/U(1) = SU(2)$ followed by π. It also follows that the fibres of H are just the orbits of the action of S^1 on S^3. In fact, another (usually preferred) way to define the Hopf map is to say that it is the orbit map of the action of the center $S^1 \subset U(2)$ on $S^3 \subset \mathbf{C}^2$. Notice that $S^2 = CP^1$ follows immediately. The advantage of this definition is that it can be generalized to higher dimensions. In fact, the orbit map

$$H : S^{2m+1} \to CP^m$$

of the action of the diagonal center $S^1 \subset U(m+1)$ on the unit sphere $S^{2m+1} \subset \mathbf{C}^{m+1}$ is also said to be the Hopf map.

REMARK: Let $\mu : S^3 \to [-1,1] \subset \mathbf{R}$ be the (isoparametric) function defined by

$$\mu(z,w) = |z|^2 - |w|^2, \ (z,w) \in S^3.$$

Then, for $|c| < 1$, the level surface

$$\mu^{-1}(c) = \sqrt{\frac{1+c}{2}} \cdot S^1 \times \sqrt{\frac{1-c}{2}} \cdot S^1 \subset S^3 \subset \mathbf{C}^2$$

is a Clifford torus. Moreover, $\mu^{-1}(\pm 1)$ are orthogonal circles that are nothing but the intersections of the components of $\mathbf{C}^2$ with S^3. Since μ is invariant under the action of S^1 on S^3, each Clifford torus is foliated by fibres of H. Actually, the restriction of the action to a Clifford torus is a periodic flow with rotation number 1. It follows that each pair of fibres (belonging to a Clifford torus) is topologically linked in S^3, in particular, $H : S^3 \to S^2$ is homotopically nontrivial.

In real coordinates $z = x + iy$ and $w = u + iv$, the Hopf map takes the form

(4.5) $\qquad H(x,y,u,v) = (x^2 + y^2 - u^2 - v^2, 2(xu + yv), 2(yu - xv)).$

28

We observe that the components of H are harmonic (homogeneous) quadratic polynomials on $\mathbf{R}^4$, or restricting to S^3, quadratic spherical harmonics on S^3, i.e. eigenfunctions of the Laplace-Beltrami operator $\triangle^{S^3}$ corresponding to the second eigenvalue $\lambda_2\ (= 8)$. Thus we have

$$\triangle^{S^3} H = \lambda_2 H$$

so that, by Proposition 1.1, the Hopf map is a λ_2-eigenmap, in particular, it is harmonic (cf. [Fuller]). By Example 2.2, $\mathcal{L}_H = \mathcal{E}_H = \{0\}$ since H is onto.

PROBLEM: Determine the (divergencefree) Jacobi fields along the Hopf map explicitly (cf. [Urakawa;1]).

The vector space $\mathcal{H}^2_{S^3}(\mathbf{R})$ of quadratic spherical harmonics on S^3 is 9-dimensional and is spanned by components of the λ_2-eigenmap

$$f_{\lambda_2} : S^3 \to S^8$$

given by

$$f_{\lambda_2}(x, y, u, v) = \sqrt{\frac{2}{3}}\left(\frac{1}{\sqrt{2}}(x^2 + y^2 - u^2 - v^2), x^2 - y^2, u^2 - v^2, \right.$$

(4.6)
$$\left. 2xy, 2uv, 2xu, 2xv, 2yu, 2yv\right).$$

Clearly, for a full harmonic map $f : S^3 \to S^n$, we have $f \leftharpoonup f_{\lambda_2}$ iff f is a λ_2-eigenmap. In this case $2 \le n \le 8$. (Note that there is no nonconstant harmonic map $f : M \to S^1$ of a compact simply connected Riemannian manifold M into the unit circle S^1 since, passing to the universal covering, f induces a harmonic function on M which must be constant by the maximum principle.) Now $H \leftharpoonup f_{\lambda_2}$ so that $\langle H \rangle_{f_{\lambda_2}}$ lies on the boundary of the moduli space $\mathcal{L}_{f_{\lambda_2}}$. With respect to the natural saturation $\mathcal{I}_{f_{\lambda_2}}$ on $\mathcal{L}_{f_{\lambda_2}}$, I_H is a one point cell consisting of $\langle H \rangle_{f_{\lambda_2}}$ alone.

Like the Hopf map, $f_{\lambda_2} : S^3 \to S^8$ is also classical. Its components form an orthonormal basis of $\mathcal{H}^2_{S^3}(\mathbf{R})$ with respect to the L^2-scalar product suitably normalized. This implies, as will be shown in the next chapter, that $f_{\lambda_2} : S^3 \to S^8$

is a minimal immersion (with induced Riemannian metric on S^3 that is a constant multiple of the standard Riemannian metric of curvature 1.) $f_{\lambda_2} : S^3 \to S^8$ is said to be the *standard minimal immersion* associated to the eigenvalue λ_2. (As it is quadratic, it factors through the natural projection $\pi : S^3 \to \mathbf{R}P^3$ yielding an imbedding of the real projective space $\mathbf{R}P^3$ into S^8 as a minimal submanifold.) Clearly, $f_{\lambda_2} : S^3 \to S^8$ is equivariant with respect to the homomorphism $\rho_{\lambda_2} : O(4) \to O(9)$ that is just the $O(4)$-module structure of $\mathcal{H}^2_{S^3}(\mathbf{R})$ given by precomposition of quadratic spherical harmonics with isometries in $O(4)$.

The principal problem of Chapter II is to classify all full harmonic maps between Euclidean spheres whose components are harmonic homogeneous polynomials of the same degree (k say) or, what is the same, to classify all full λ_k-eigenmaps, where λ_k is the k-th eigenvalue of the Laplace-Beltrami operator on a Euclidean sphere (of arbitrary dimension).

In our special case at hand, this boils down to the understanding of the moduli space $\mathcal{L}_{f_{\lambda_2}}$. While the full geometric description of $\mathcal{L}_{f_{\lambda_2}}$ will only be given later (due primarily to the fact that $\dim \mathcal{L}_{f_{\lambda_2}} = 10$, as will be shown in the sequel), we see no harm in exploring some part of the boundary $\partial \mathcal{L}_{f_{\lambda_2}}$.

THEOREM 4.1. *(a) Let $f : S^3 \to S^2$ be a full λ_2-eigenmap. Then there exist $U \in O(3)$ and $a \in O(4)$ such that*

$$f = U \circ H \circ a.$$

(b) There is no full λ_2-eigenmap $f : S^3 \to S^3$.

REMARK: In the following (admittedly long) proof it will be shown that, with respect to the action of $O(4)$ on $\mathcal{L}_{f_{\lambda_2}}$, the orbit $O(4) \cdot \langle H \rangle_{f_{\lambda_2}}$ consists of two copies $\mathbf{R}P^2_1$ and $\mathbf{R}P^2_2$ of the real projective plane imbedded in the boundary of $\mathcal{L}_{f_{\lambda_2}}$. In the next chapter we will prove that $\mathcal{E}_{f_{\lambda_2}}$ has two 5-dimensional irreducible $SO(4)$-components V_1 and V_2 (by restriction) and, for $l = 1, 2$, the $SO(4)$-orbit $\mathbf{R}P^2_l$ sits in

S_{V_l} as the classical (minimal) Veronese surface $\mathbf{R}P^2$ in S^4. This latter imbedding is, of course, factored from the standard minimal immersion $f_{\lambda_2} : S^2 \to S^4$ known as the real Veronese map.

PROBLEM: Show that if $f : S^2 \to S^n$ is a full λ_2-eigenmap then f is a standard minimal immersion, i.e. $n = 4$ and the components of f form an orthonormal basis in the space $\mathcal{H}_{S^2}^2(\mathbf{R})$ of quadratic spherical harmonics (with respect to the L^2-scalar product suitably normalized). Work out an explicit expression for a standard minimal immersion $f_{\lambda_2} : S^2 \to S^4$. Show that f_{λ_2} factors through the projection $S^2 \to \mathbf{R}P^2$ inducing an imbedding of $\mathbf{R}P^2$ into S^4 (cf. [Berger-Gauduchon-Mazet]).

PROOF OF THEOREM 4.1: Let $f : S^3 \to S^n$ be a full λ_2-eigenmap. Using real coordinates $(x, y, u, v) \in \mathbf{R}^4$ we write

$$f(x, y, u, v) = a_{11}x^2 + a_{22}y^2 + a_{33}u^2 + a_{44}v^2$$
$$+ a_{12}xy + a_{13}xu + a_{14}xv + a_{23}yu + a_{24}yv + a_{34}uv,$$

where the vectors $a_{ij} \in \mathbf{R}^{n+1}$, $1 \le i \le j \le 4$, span $\mathbf{R}^{n+1}$ and, by harmonicity

(4.7) $$\sum_{i=1}^{4} a_{ii} = 0.$$

In what follows, to simplify the notation, we put $a_i = a_{ii}$, $1 \le i \le 4$. Moreover, every double index below will be thought to be symmetric, in particular, $a_{ij} = a_{ji}$, $1 \le i, j \le 4$.

The condition that f maps into S^n can, by homogeneity, be rephrased as

(4.8) $$|f(x, y, u, v)|^2 = (x^2 + y^2 + u^2 + v^2)^2$$

that has to be satisfied *for all* $(x, y, u, v) \in \mathbf{R}^4$. This is because the left hand side of (4.8) is a homogeneous polynomial of degree 4 on $\mathbf{R}^4$. Expanding both sides of

(4.8), we obtain the following orthogonality relations

$$|a_i|^2 = 1, \ 1 \le i \le 4,$$

$$\langle a_i, a_{ij} \rangle = 0, \ 1 \le i < j \le 4,$$

(4.9) $$|a_{ij}|^2 + 2\langle a_i, a_j \rangle = 2, \ 1 \le i < j \le 4,$$

$$\langle a_i, a_{jk} \rangle + \langle a_{ij}, a_{ik} \rangle = 0, \ 1 \le i, j, k \le 4, \ i, j, k \quad \text{distinct},$$

$$\langle a_{ij}, a_{kl} \rangle + \langle a_{ik}, a_{jl} \rangle + \langle a_{il}, a_{jk} \rangle = 0, \ 1 \le i, j, k, l \le 4, \ i, j, k, l \quad \text{distinct}.$$

A system $\{a_{ij}\}_{i,j=0}^4 \subset \mathbf{R}^{n+1}$ of vectors is said to be *feasible* if it spans $\mathbf{R}^{n+1}$, satisfies (4.7), and the orthogonality relations (4.9). Two feasible systems are said to be *orthogonally equivalent* if there exists a linear isometry $U \in O(n+1)$ carrying one system to the other. In terms of the corresponding harmonic maps, this is just the relation of orthogonal equivalence.

The problem is now a linear algebraic one: Determine (or parametrize) the set of orthogonal equivalence classes of feasible systems.

To bring geometry in, first consider only the subsystem $\{a_i\}_{i=1}^4 \subset \mathbf{R}^{n+1}$. Multiplying (4.7) by a_j, $1 \le j \le 4$, and using the first orthogonality relation in (4.9), we obtain

$$\langle a_1, a_2 \rangle = \langle a_3, a_4 \rangle = \cos \alpha,$$

$$\langle a_1, a_3 \rangle = \langle a_2, a_4 \rangle = \cos \beta,$$

$$\langle a_1, a_4 \rangle = \langle a_2, a_3 \rangle = \cos \gamma,$$

where the angles $0 \le \alpha, \beta, \gamma \le \pi$ satisfy

$$\cos \alpha + \cos \beta + \cos \gamma = -1.$$

For $0 < \alpha < \pi$, we think of $\{a_i\}_{i=1}^4$ as a *bowtie* consisting of two congruent isosceles triangles Δ_1 and Δ_2 that are spanned by $\{a_1, a_2\}$ and $\{a_3, a_4\}$, respectively. The

32

angle of Δ_l, $l = 1, 2$, at the origin is α. By (4.7), $\dim \operatorname{span}\{a_i\}_{i=1}^4 \leq 3$. When equality holds, we call the bowtie *twisted*; the angle δ at which (the planes spanned by) Δ_1 and Δ_2 are tilted is given by

$$\cos\delta = \langle a_1 - a_2, a_3 - a_4 \rangle = 2(\cos\beta - \cos\gamma).$$

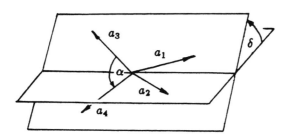

From here on the computations are elementary but amazingly long so that we give only somewhat sketchy details.

Let $n = 2$. We assume that we have a twisted bowtie since the other cases will be obtained by letting $\alpha \to 0, \pi$ and $\delta \to 0, \pi$. Now somewhat tedious computation shows that for a fixed twisted bowtie $\{a_i\}_{i=1}^4 \subset \mathbf{R}^3$ the rest of the vectors in a feasible system are given by

$$a_{ij} = \varepsilon_{ij}\sqrt{\frac{2}{1 + \langle a_i, a_j \rangle}} \cdot a_i \times a_j, \quad 1 \leq i < j \leq 4,$$

where $\times$ is the vector cross product in $\mathbf{R}^3$ and $\varepsilon_{ij} = \pm 1$. Moreover, the signs ε_{ij} (with symmetric indices) satisfy

(4.10) $$\varepsilon_{ij}\varepsilon_{ik} = (-1)^{j+k}\varepsilon_{c(ij)}\varepsilon_{c(ik)},$$

for any triple of distinct indices i, j, k, where c stands for the complement with respect to the set $\{1, 2, 3, 4\}$. Thus any position of the twisted bowtie and any feasible set of signs $\varepsilon = \{\varepsilon_{ij}\}_{i,j=1}^4 \subset \mathbf{Z}^6$ (i.e. which satisfies (4.10)) determine a full

33

harmonic map $f : S^3 \to S^2$ derived from f_{λ_2}. We now fix a feasible $\varepsilon \in \mathbf{Z}^6$. We also place the twisted bowtie in the natural position given as

(4.11)

$$a_1 = (\cos\frac{\alpha}{2}, \sin\frac{\alpha}{2}, 0),$$

$$a_2 = (\cos\frac{\alpha}{2}, -\sin\frac{\alpha}{2}, 0),$$

$$a_3 = (-\cos\frac{\alpha}{2}, -\cos\delta\sin\frac{\alpha}{2}, -\sin\delta\sin\frac{\alpha}{2}),$$

$$a_4 = (-\cos\frac{\alpha}{2}, \cos\delta\sin\frac{\alpha}{2}, \sin\delta\sin\frac{\alpha}{2}).$$

The full harmonic map $f^\varepsilon_{\alpha,\delta} : S^3 \to S^2$ that corresponds to these data is

$$
\begin{aligned}
f^\varepsilon_{\alpha,\delta}(x,y,u,v) = \Big(&\cos\frac{\alpha}{2}(x^2+y^2-u^2-v^2) + 2\sin\frac{\alpha}{2}\sin\frac{\delta}{2}(\varepsilon_3 xv + \varepsilon_4 yu)\\
&- 2\sin\frac{\alpha}{2}\cos\frac{\delta}{2}(\varepsilon_2 xu + \varepsilon_5 yv),\\
&\sin\frac{\alpha}{2}(x^2-y^2-\cos\delta(u^2-v^2)) + 2\cos\frac{\alpha}{2}\cos\frac{\delta}{2}(\varepsilon_2 xu - \varepsilon_5 yv)\\
&- 2\cos\frac{\alpha}{2}\sin\frac{\delta}{2}(\varepsilon_3 xv - \varepsilon_4 yu) + 2\varepsilon_6\sin\frac{\alpha}{2}\sin\delta uv,\\
&- \sin\frac{\alpha}{2}\sin\delta(u^2-v^2) - 2\sin\frac{\alpha}{2}(\varepsilon_1 xy + \varepsilon_6\cos\delta uv)\\
&+ 2\cos\frac{\alpha}{2}\sin\frac{\delta}{2}(\varepsilon_2 xu - \varepsilon_5 yv) + 2\cos\frac{\alpha}{2}\cos\frac{\delta}{2}(\varepsilon_3 xv - \varepsilon_4 yu)\Big).
\end{aligned}
$$

We now simply substitute $\alpha, \delta = 0, \pi$ to obtain the limiting cases. As easily checked, $f^\varepsilon_{\alpha,\delta}$ is orthogonally equivalent to $f^\varepsilon_{\alpha',\delta'}$ iff $\alpha = \alpha' = 0$. For fixed feasible $\varepsilon \in \mathbf{Z}^6$, we then obtain that the closed triangle

$$[0,\pi] \times [0,\pi]/\{0\} \times [0,\pi]$$

is imbedded into $\partial\mathcal{L}_{f_{\lambda_2}}$ by the parametrization (2.5). Now the sign relations (4.10) boil down to eight feasible solutions that correspond to eight imbedded triangles in $\partial\mathcal{L}_{f_{\lambda_2}}$. Full eigenmaps corresponding to interior points of distinct triangles are orthogonally inequivalent. There are various orthogonal equivalences however between full eigenmaps corresponding to boundary points. In $\partial\mathcal{L}_{f_{\lambda_2}}$ this means that

34

the eight triangles are pasted together along their edges. Inspection shows that they form two copies of the real projective plane.

On the other hand, as noted above, the Hopf map $H : S^3 \to S^2$ is equivariant with respect to a homomorphism $\rho_H : U(2) \to O(3)$. In terms of the moduli space $\mathcal{L}_{f_{\lambda_2}}$, this means that the unitary group $U(2)$ is contained in the isotropy subgroup of the action of $O(4)$ on $\mathcal{L}_{f_{\lambda_2}}$ at $\langle H \rangle_{f_{\lambda_2}}$. Moreover, there is no closed subgroup between $U(2)$ and $SO(4)$ and so the former actually coincides with the isotropy subgroup. Hence the $O(4)$-orbit $\langle H \rangle_{f_{\lambda_2}}$ is topologically $O(4)/U(2)$. This homogeneous space however consists of two copies of the real projective plane so that it must coincide with what we obtained previously. The proof of part (a) is complete.

Now let $n = 3$. We fix a twisted bowtie given by (4.11) that spans $\mathbf{R}^3 = \mathbf{R}^3 \times \{0\} \subset \mathbf{R}^4$. Putting $e = (0,0,0,1) \in \mathbf{R}^4$, we decompose

$$a_{ij} = a'_{ij} + \lambda_{ij}e,\ 1 \leq i < j \leq 4,$$

where $a'_{ij} \in \mathbf{R}^3$ and $\lambda_{ij} \in \mathbf{R}$. Computation in the use of some of the orthogonality relations in (4.9) then shows that

$$a'_{ij} = \varepsilon_{ij}\sqrt{\frac{2(1 - \langle a_i, a_j \rangle) + \lambda_{ij}^2}{1 - \langle a_i, a_j \rangle^2}}a_i \times a_j,\ 1 \leq i < j \leq 4.$$

Substituting these into (4.9), it follows that $\lambda_{ij} = 0$ for all $1 \leq i < j \leq 4$ which contradicts to fullness. Finally the limiting cases can be settled by case-by-case verification. The proof of (b) is complete. $\sqrt{}$

REMARK: Quaternionic squaring is a full map $f : S^3 \to S^3$ whose components are homogeneous quadratic polynomials on $\mathbf{R}^4$. Clearly, f is not harmonic. In general, if n is odd and $k \in \pi_n(S^n) \cong \mathbf{Z}$ then k can be represented by a polynomial map $f : S^n \to S^n$ of degree $|k|$. Moreover, if n is a power of 2 then all polynomial maps $f : S^n \to S^{n-1}$ are constant (cf. [R.Wood]).

The components of the complex form (4.3) of the Hopf map are harmonic homogeneous polynomials on $\mathbf{C}^2$ of bidegree (1,1), i.e. of degree one in z, w and degree one in the conjugates $\bar{z}, \bar{w}$. Let $\mathcal{H}^{1,1}_{\mathbf{C}P^1}$ denote the *complex* vector space of complex valued harmonic homogeneous polynomials of bidegree (1,1) on $\mathbf{C}^2$. The subscript $\mathbf{C}P^1$ indicates that a polynomial in $\mathcal{H}^{1,1}_{\mathbf{C}P^1}$ is uniquely detemined by its values taken on $\mathbf{C}P^1$. Note that the components of the Hopf map H in (4.3) form an orthonormal basis in $\mathcal{H}^{1,1}_{\mathbf{C}P^1}$ with respect to the Hermitian L^2-scalar product (defined by integration on S^3) suitably normalized. $\mathcal{H}^{1,1}_{\mathbf{C}P^1}$ is a complex irreducible $U(2)$-module, where the module structure is given by precomposition of polynomials with unitary transformations in $U(2)$ which clearly leave the bidegree invariant. To put this into a representation theoretical framework, we view $\mathcal{H}^{1,1}_{\mathbf{C}P^1}$ as a $U(2)$-component of the complexification $\mathcal{H}^2_{S^3}$ of $\mathcal{H}^2_{S^3}(\mathbf{R})$ by restriction from $O(4)$ to $U(2)$. In fact, the complete decomposition of the $O(4)$-module of real quadratic spherical harmonics on S^3 is given as

$$\mathcal{H}^2_{S^3}\big|_{U(2)} \cong \mathcal{H}^{2,0}_{\mathbf{C}P^1} \oplus \mathcal{H}^{1,1}_{\mathbf{C}P^1} \oplus \mathcal{H}^{0,2}_{\mathbf{C}P^1}.$$

Here $\mathcal{H}^{p,q}_{\mathbf{C}P^1}$ denotes the complex irreducible $U(2)$-module of (complex valued) harmonic homogeneous polynomials on $\mathbf{C}^2$ of degree p in z, w and degree q in $\bar{z}, \bar{w}$. Note that $\mathcal{H}^{0,2}_{\mathbf{C}P^1}$ is just the dual of $\mathcal{H}^{2,0}_{\mathbf{C}P^1}$. Based on the analogy with the Hopf map, we wish to construct a harmonic map by taking its components from an orthonormal basis of $\mathcal{H}^{2,0}_{\mathbf{C}P^1}$ (again with respect to the Hermitian L^2-scalar product suitably normalized). We then arrive at the classical *complex Veronese map*

$$V : S^3 \rightarrow S^5$$

defined by

(4.12) $$V(z, w) = (z^2, \sqrt{2}zw, w^2).$$

36

The complex Veronese map is clearly equivariant with respect to the homomorphism $\rho_V : U(2) \to U(3) \subset O(6)$ given by the unitary module structure of $\mathcal{H}_{\mathbf{C}P^1}^{2,0}$. Geometrically, and classically, V factors through the Hopf maps $H : S^3 \to \mathbf{C}P^1$ and $H : S^5 \to \mathbf{C}P^2$ yielding the map

$$V : \mathbf{C}P^1 \to \mathbf{C}P^2$$

which by abuse of terminology and respect of history is also called the complex Veronese map. Actually, V is a holomorphic imbedding of $\mathbf{C}P^1$ into $\mathbf{C}P^2$ and the image of V is the complex surface $2XZ = Y^2$ in homogeneous coordinates $[X : Y : Z]$ on $\mathbf{C}P^2$.

In real coordinates, we have

$$V(x, y, u, v) = (x^2 - y^2, 2xy, \sqrt{2}(xu - yv), \sqrt{2}(xv + yu), u^2 - v^2, 2uv).$$

We wish to determine $\mathcal{E}_V$. Computation shows that

$$\mathcal{E}_V = \{C(a, b) \in S^2(\mathbf{R}^6) | a, b \in \mathbf{R}\},$$

where $C(a, b)$ has the form

$$
\begin{bmatrix}
0 & 0 & 0 & 0 & a & -b \\
0 & 0 & 0 & 0 & -b & -a \\
0 & 0 & -a & b & 0 & 0 \\
0 & 0 & b & a & 0 & 0 \\
a & -b & 0 & 0 & 0 & 0 \\
-b & -a & 0 & 0 & 0 & 0
\end{bmatrix}
$$

Now $C(a, b) + I_6 \geq 0$ iff $a^2 + b^2 \leq 1$ so that $\mathcal{L}_V \subset \mathcal{E}_V$ is a 2-dimensional disk of radius 6 in $S^2(\mathbf{R}^6)$ with respect to the scalar product (2.1). To describe the boundary $\partial \mathcal{L}_V$ we first define

$$H_\alpha : S^3 \to S^2$$

37

in complex coordinates by

$$(4.13) \qquad H_\alpha = (e^{2i\alpha}z^2 + \bar{w}^2, 2\operatorname{Im}(e^{i\alpha}zw)).$$

Writing H_α out in real coordinates, we infer that $H_\alpha \dashv V$ so that $\{\langle H_\alpha \rangle_V | \alpha \in \mathbf{R}\} \subset \partial \mathcal{L}_V$. Moreover, for $\alpha_1 \not\equiv \alpha_2 \,(\operatorname{mod} \pi)$, H_{α_1} and H_{α_2} are orthogonally inequivalent so that actually $\{\langle H_\alpha \rangle_V | \alpha \in \mathbf{R}\} = \partial \mathcal{L}_V$. By (the proof of) Theorem 4.1, $\mathcal{L}_V$ is a 2-disk whose boundary circle lies in a copy of the real projective plane $\mathbf{R}P^2$. As for the $U(2)$-action on $\mathcal{L}_V$ induced by the equivariance of V, the center S^1 given in (4.4) acts as

$$(4.14) \qquad \operatorname{diag}(e^{i\theta}, e^{i\theta}) \cdot \langle H_\alpha \rangle_V = \langle H_{\alpha+2\theta} \rangle_V$$

so that S^1 rotates the 2-disk $\mathcal{L}_V$. $SU(2) = U(2)/S^1$ acts trivially on $\mathcal{L}_V$.

§5. Direct product

Let M_1 and M_2 be Riemannian manifolds and assume that λ is a common eigenvalue of the Laplace-Beltrami operators $\triangle^{M_1}$ and $\triangle^{M_2}$ acting on functions on M_1 and M_2, respectively. Let $f_1 : M_1 \to S_{V_1}$ and $f_2 : M_2 \to S_{V_2}$ be full λ-eigenmaps, where V_1 and V_2 are Euclidean vector spaces. By definition, for $l = 1, 2$, we have

$$(5.1) \qquad \triangle^{M_l} f_l = \lambda \cdot f_l.$$

We define

$$f_1 \times f_2 : M_1 \times M_2 \to S_{V_1 \times V_2}$$

by

$$(f_1 \times f_2)(x, y) = \frac{1}{\sqrt{2}}(f_1(x), f_2(y)), \ x \in M_1, \ y \in M_2.$$

38

By (5.1), we have

$$\triangle^{M_1 \times M_2}(f_1 \times f_2) = \lambda \cdot (f_1 \times f_2)$$

so that $f_1 \times f_2 : M_1 \times M_2 \to S_{V_1 \times V_2}$ is a full λ-eigenmap called the *direct product* of f_1 and f_2.

The definition clearly generalizes to direct product of finitely many λ-eigenmaps.

THEOREM 5.1. *There is a linear isomorphism*

$$\Phi : \mathcal{E}_{f_1 \times f_2} \to \mathcal{E}_{f_1} \times \mathcal{E}_{f_2} \times \mathbf{R}$$

under which the moduli space $\mathcal{L}_{f_1 \times f_2} \subset \mathcal{E}_{f_1 \times f_2}$ *corresponds to the convex hull of the sets*

$$2\mathcal{L}_{f_1} \times \{0\} \times \{1\} \qquad and \qquad \{0\} \times 2\mathcal{L}_{f_2} \times \{-1\}$$

on $\mathcal{E}_{f_1} \times \mathcal{E}_{f_2} \times \mathbf{R}$.

PROOF: Define

$$\Psi : \mathcal{E}_{f_1} \times \mathcal{E}_{f_2} \times \mathbf{R} \to S^2(V_1 \times V_2)$$

by

$$\Psi(C_1, C_2, c) = (C_1 + cI_{V_1}) \times (C_2 - cI_{V_2}), \; C_l \in \mathcal{E}_{f_l}, \, l = 1, 2, \, c \in \mathbf{R}.$$

We have

$$\langle \Psi(C_1, C_2, c) \circ (f_1 \times f_2), (f_1 \times f_2) \rangle = \frac{1}{2}\langle (C_1 + cI_{V_1}) \circ f_1, f_1 \rangle + \frac{1}{2}\langle (C_2 - cI_{V_2}) \circ f_2, f_2 \rangle$$
$$= \frac{1}{2}\langle C_1 \circ f_1, f_1 \rangle + \frac{c}{2} + \frac{1}{2}\langle C_2 \circ f_2, f_2 \rangle - \frac{c}{2} = 0$$

so that Ψ maps into $\mathcal{E}_{f_1 \times f_2}$. Restricting the image, we claim that

$$\Psi : \mathcal{E}_{f_1} \times \mathcal{E}_{f_2} \times \mathbf{R} \to \mathcal{E}_{f_1 \times f_2}$$

is an isomorphism.

39

First we show that Ψ is injective. Let $C_1 + cI_{V_1} = 0$ and $C_2 - cI_{V_2} = 0$ for some $C_l \in \mathcal{E}_{f_l}$, $l = 1, 2$, and $c \in \mathbf{R}$. From the first equality we obtain

$$c = c|f_1|^2 + \langle C_1, \mathrm{proj}\,[f_1]\rangle = \langle (C_1 + cI_{V_1}) \circ f_1, f_1 \rangle = 0$$

so that $C_1 = 0$ follows. Since $c = 0$ the second equality implies $C_2 = 0$.

Secondly, we show that Ψ maps onto $\mathcal{E}_{f_1 \times f_2}$. Let $C \in \mathcal{E}_{f_1 \times f_2}$ and write C in the form

$$\begin{bmatrix} C_1 & B^{\mathsf{T}} \\ B & C_2 \end{bmatrix}$$

where $C_l \in S^2(V_l)$ and $B : V_1 \to V_2$ is a linear map. With these notations, we have (on $M_1 \times M_2$)

(5.2) $\quad \langle C \circ (f_1 \times f_2), (f_1 \times f_2) \rangle = \langle C_1 \circ f_1, f_1 \rangle + 2\langle B \circ f_1, f_2 \rangle + \langle C_2 \circ f_2, f_2 \rangle = 0.$

Applying $\triangle^{M_1} \cdot \triangle^{M_2}$ to both sides and using (5.1), we obtain

$$\langle B \circ f_1, f_2 \rangle = 0.$$

By fullness $B = 0$ follows and hence (5.2) reduces to

$$\langle C_1 \circ f_1, f_1 \rangle + \langle C_2 \circ f_2, f_2 \rangle = 0.$$

The l-th term on the left hand side is a function on M_l, $l = 1, 2$, so that

$$\langle C_1 \circ f_1, f_1 \rangle = -\langle C_2 \circ f_2, f_2 \rangle$$

is a constant, c say. This however means that

$$C_1 - cI_{V_1} \in \mathcal{E}_{f_1} \qquad \text{and} \qquad C_2 + cI_{V_2} \in \mathcal{E}_{f_2}.$$

Working backwards, we have $\Psi(C_1 - cI_{V_1}, C_2 + cI_{V_2}, c) = C$ and Ψ is onto. Setting $\Phi = \Psi^{-1}$, we obtain the first statement.

40

To prove the second, we first note that, for $(C_1, C_2, c) \in \mathcal{E}_{f_1} \times \mathcal{E}_{f_2} \times \mathbf{R}$, the condition $\Psi(C_1, C_2, c) \in \mathcal{L}_f$ holds iff

(5.3) $\qquad C_1 + (1+c)I_{V_1} \geq 0 \qquad$ and $\qquad C_2 + (1-c)I_{V_2} \geq 0.$

Clearly, $|c| \leq 1$ since otherwise $C_1 > 0$ or $C_2 > 0$ so that $\langle C_1 \circ f_1, f_1 \rangle = 0$ or $\langle C_2 \circ f_2, f_2 \rangle = 0$ would be violated. For $c = 1$, we have $\frac{1}{2}C_1 + I_{V_1} \geq 0$ and $C_2 = 0$ and these make up $2\mathcal{L}_{f_1} \times \{0\} \times \{1\}$. For $c = -1$, we get $\frac{1}{2}C_2 + I_{V_2} \geq 0$ and $C_1 = 0$ arriving at $\{0\} \times 2\mathcal{L}_{f_2} \times \{-1\}$. The rest is clear by (5.3) and convexity. $\checkmark$

THEOREM 5.2. Let $f_l : M_l \rightarrow S_{V_l}$, $l = 1, 2$, be minimal immersions. Then $f_1 \times f_2 :$ $M_1 \times M_2 \rightarrow S_{V_1 \times V_2}$ is a minimal immersion and

$$M_{f_1 \times f_2} = M_{f_1} \times M_{f_2} \times \{0\} \subset \mathcal{L}_{f_1 \times f_2},$$

under the isomorphism of Theorem 5.1.

PROOF: By Theorem 2.7, we have

$$\mathcal{F}_{f_1 \times f_2} \subset \mathcal{E}_{f_1 \times f_2} \cong \mathcal{E}_{f_1} \times \mathcal{E}_{f_2} \times \mathbf{R}$$

so that an element C in $\mathcal{F}_{f_1 \times f_2}$ can be written as $(C_1 + cI_{V_1}, C_2 - cI_{V_2})$, where $C_l \in \mathcal{E}_{f_l}$, $l = 1, 2$, and $c \in \mathbf{R}$. Now, for $X_x \in T_x(M_1)$ and $Y_y \in T_y(M_2)$, we have

$$\langle (C_1 + cI_{V_1})(f_1)_* X_x \check{\ }, (f_1)_* X_x \check{\ } \rangle + \langle (C_2 - cI_{V_2})(f_2)_* Y_y \check{\ }, (f_2)_* Y_y \check{\ } \rangle = 0$$

or equivalently

$$\langle C_1 (f_1)_* X_x \check{\ }, (f_1)_* X_x \check{\ } \rangle + c|X_x|^2 + \langle C_2 (f_2)_* Y_y \check{\ }, (f_2)_* Y_y \check{\ } \rangle - c|Y_y|^2 = 0.$$

Since the lengths of X_x and Y_y can be chosen arbitrarily, we obtain

$$\langle C_1 (f_1)_* X_x \check{\ }, (f_1)_* X_x \check{\ } \rangle + c|X_x|^2 = 0$$

41

so that $C_1 + cI_{V_1} \in \mathcal{F}_{f_1} \subset \mathcal{E}_{f_1}$. On the other hand $C_1 \in \mathcal{E}_{f_1}$. Hence $cI_{V_1} \in \mathcal{E}_{f_1}$ which is possible only if $c = 0$. Summarizing, we obtained that

$$\mathcal{M}_{f_1 \times f_2} \subset \mathcal{M}_{f_1} \times \mathcal{M}_{f_2} \times \{0\}.$$

The reverse inclusion is obvious. $\checkmark$

EXAMPLE 5.3: Let V_j, $j = 1, \ldots, k$ be Euclidean vector spaces and consider the linear imbedding

$$f = \prod_{j=1}^{k} I_{S_{V_j}} : \prod_{j=1}^{k} S_{V_j} \to S_V, \, V = \prod_{j=0}^{k} V_j$$

given by

$$f(x_1, \ldots, x_k) = \frac{1}{\sqrt{k}}(x_1, \ldots, x_k), \, x_j \in S_{V_j}, \, j = 1, \ldots, k.$$

By (the proof of) Theorem 5.1, we have

$$\mathcal{E}_f = \{\prod_{j=1}^{k}(c_j I_{V_j}) | \sum_{j=1}^{k} c_j = 0\} \cong \mathbf{R}^{k-1}.$$

Moreover, the moduli space of f is the $(k-1)$-simplex

$$\mathcal{L}_f = \{\prod_{j=1}^{k}(c_j I_{V_j}) | \sum_{j=1}^{k} c_j = 0 \text{ and } |c_j| \le 1, \, j = 1, \ldots, k \}.$$

Since f is minimal (with induced metric a constant multiple of the product metric) $\mathcal{F}_f$ exists and, by Theorem 5.2, consists of the origin only. Note that, for $|c_j| < 1, j = 1, \ldots, k$, the corresponding eigenmap is also minimal but the induced metric is different from the one induced by f provided that $(c_1, \ldots, c_k)$ is not the origin. For $k = 2$,

$$\mathcal{L}_f = [-1, 1] = \{ \text{diag}\,(aI_{V_1}, -aI_{V_2}) | \, |a| \le 1\}.$$

For $s \in \mathbf{R}$, let

$$f_s : S_{V_1} \times S_{V_2} \to S_{V_1 \times V_2}$$

be given by

$$f_s(x, y) = (\cos s \, x, \sin s \, y), \, x \in S_{V_1}, \, y \in S_{V_2}.$$

Then $\langle f_s \rangle_f = \operatorname{diag}(2\cos^2 s\, I_{V_1}, 2\sin^2 s\, I_{V_2}) - I_V = \operatorname{diag}(\cos(2s)I_{V_1}, \sin(2s)I_{V_2})$ and it corresponds to $\operatorname{diag}(aI_{V_1}, -aI_{V_2}) \in \mathcal{L}_f$. Note that the images of these maps (that comprise the whole moduli space $\mathcal{L}_f$) form what is called an isoparametric family of hypersurfaces of degree 2 on S_V. In particular, for $V_1 = V_2 = \mathbf{R}^2$, we recover the Clifford tori in S^3 discussed above.

§6. Tensor product

Let M_1 and M_2 be Riemannian manifolds and $f_1 : M_1 \to S_{V_1}$ and $f_2 : M_2 \to S_{V_2}$ full harmonic maps, where V_1 and V_2 are Euclidean vector spaces. By Proposition 1.1, we have

$$(6.1) \qquad \triangle^{M_l} f_l = e(f_l) \cdot f_l, \; l = 1, 2.$$

We define the *tensor product*

$$f_1 \otimes f_2 : M_1 \times M_2 \to S_{V_1 \otimes V_2}$$

of f_1 and f_2 by

$$(f_1 \otimes f_2)(x, y) = f_1(x) \otimes f_2(y), \; x \in M_1, y \in M_2.$$

Using (6.1), we obtain

$$\triangle^{M_1 \times M_2}(f_1 \otimes f_2) = (e(f_1) + e(f_2))f_1 \otimes f_2$$

so that the tensor product $f_1 \otimes f_2 : M_1 \times M_2 \to S_{V_1 \otimes V_2}$ is a full harmonic map with energy density

$$e(f_1 \otimes f_2) = e(f_1) + e(f_2).$$

As for the direct product, the definition clearly extends to the tensor product of finitely many harmonic maps.

THEOREM 6.1. *We have*

(6.2) $\mathcal{E}_{f_1 \otimes f_2} = so(V_1) \otimes so(V_2) + \mathcal{E}_{f_1} \otimes S^2(V_2) + S^2(V_1) \otimes \mathcal{E}_{f_2}.$

PROOF: Clearly, every term on the right hand side of (6.2) is a linear subspace of $\mathcal{E}_{f_1 \otimes f_2} \subset S^2(V_1 \otimes V_2)$. To prove the reverse inclusion, let $C \in \mathcal{E}_{f_1 \otimes f_2}$. In general, we have

$$S^2(V_1 \otimes V_2) = so(V_1) \otimes so(V_2) + S^2(V_1) \otimes S^2(V_2)$$

so that we can write

(6.3) $$C = \sum_{i=1}^{r} B_1^i \otimes B_2^i + \sum_{j=1}^{s} C_1^j \otimes C_2^j,$$

where $B_l^i \in so(V_l)$ and $C_l^j \in S^2(V_l)$, $l = 1, 2$. For fixed j, we decompose

(6.4) $$C_1^j = C_1'^{\,j} + C_1''^{\,j},$$

where $C_1'^{\,j} \in \mathcal{E}_{f_1}$ and $C_1''^{\,j} \in \mathcal{E}_{f_1}^\perp$. Substituting (6.4) into (6.3), we obtain

$$C = \sum_{i=1}^{r} B_1^i \otimes B_2^i + \sum_{j=1}^{s} C_1'^{\,j} \otimes C_2^j + \sum_{j=1}^{s} C_1''^{\,j} \otimes C_2^j.$$

As $B_1^i \otimes B_2^i \in so(V_1) \otimes so(V_2)$ and $C_1'^{\,j} \otimes C_2^j \in \mathcal{E}_{f_1} \otimes S^2(V_2)$, it remains to prove that

(6.5) $$\sum_{j=1}^{s} C_1''^{\,j} \otimes C_2^j \in S^2(V_1) \otimes \mathcal{E}_{f_2}.$$

Without loss of generality, we may assume that $\{C_1''^{\,j}\}_{j=1}^{s} \subset \mathcal{E}_{f_1}^\perp$ is a linearly independent set. $C \in \mathcal{E}_{f_1 \otimes f_2}$ translates into

(6.6) $$\sum_{j=1}^{r} \langle (C_1''^{\,j} \otimes C_2^j)(f_1 \otimes f_2), f_1 \otimes f_2 \rangle = 0.$$

This holds because B_l^i is skew and $C_1'^{\,j} \in \mathcal{E}_{f_1}$. Rewriting (6.6) as

$$\langle \sum_{j=1}^{s} \langle C_2^j f_2(y), f_2(y) \rangle C_1''^{\,j} f_1(x), f_1(x) \rangle = 0, \ x \in M_1, \ y \in M_2,$$

44

we obtain

$$\sum_{j=1}^{s} \langle C_2^j f_2(y), f_2(y) \rangle C_1''^j \in \mathcal{E}_{f_1}$$

for every $y \in M_2$. Since $C_1''^j \in \mathcal{E}_{f_1}^{\perp}$, linear independence implies that

$$\langle C_2^j f_2(y), f_2(y) \rangle = 0, \; j = 1, \ldots, s,$$

for all $y \in M_2$ so that $C_2^j \in \mathcal{E}_{f_2}$ for all $j = 1, \ldots, r$. (6.5) now follows and the proof is complete. $\checkmark$

The structure of the moduli space $\mathcal{L}_{f_1 \otimes f_2}$ in terms of $\mathcal{L}_{f_1}$ and $\mathcal{L}_{f_2}$ is fairly complicated in general. In what follows, we give an account on the special case when $f_l = I_{S_{V_l}}$, $l = 1, 2$.

Recall that an *orthogonal multiplication* on $V_1 \times V_2$ is a bilinear map $F : V_1 \times V_2 \to W$ into a Euclidean vector space W satisfying

(6.7)
$$|F(x, y)| = |x| \cdot |y|$$

for all $x \in V_1$ and $y \in V_2$. An orthogonal multiplication is said to be *full* if it is surjective. Two orthogonal multiplications $F_1 : V_1 \times V_2 \to W_1$ and $F_2 : V_1 \times V_2 \to W_2$ are said to be *orthogonally equivalent* if there exists a linear isometry $U : W_1 \to W_2$ such that $F_2 = U \cdot F_1$.

REMARK: The existence of an orthogonal multiplication

$$F : \mathbf{R}^{m+1} \times \mathbf{R}^{n+1} \to \mathbf{R}^{n+1}$$

implies the existence of m (pointwise) linearly independent vector fields on S^n. Orthogonal multiplications are also intimately related to Clifford modules that can be exploited to determine the maximal number of linearly independent vector fields on spheres (cf. [Husemoller]).

45

EXAMPLES 6.2: (a) Complex multiplication can be thought of as a full orthogonal multiplication

$$F : \mathbf{R}^2 \times \mathbf{R}^2 \rightarrow \mathbf{R}^2.$$

Similarly, quaternionic multiplication is a full orthogonal multiplication

$$F : \mathbf{R}^4 \times \mathbf{R}^4 \rightarrow \mathbf{R}^4.$$

(b) Tensor product is a full orthogonal multiplication

$$\otimes : V_1 \times V_2 \rightarrow V_1 \otimes V_2.$$

THEOREM 6.3. $\mathcal{L}_{Is_{V_1} \otimes Is_{V_2}} \subset \mathcal{E}_{Is_{V_1} \otimes Is_{V_2}} = so(V_1) \otimes so(V_2)$ *parametrizes the equivalence classes of full orthogonal multiplications on* $V_1 \times V_2$.

PROOF: Let $F : V_1 \times V_2 \rightarrow W$ be a full orthogonal multiplication. By the universal property of the tensor product, there exists a surjective linear map $A : V_1 \otimes V_2 \rightarrow W$ such that $F = A \cdot \otimes$, where $\otimes : V_1 \times V_2 \rightarrow V_1 \otimes V_2$ is the universal projection. We associate to F the symmetric endomorphism

(6.8) $$\langle F \rangle_\otimes = A^\top \cdot A - I \in S^2(V_1 \otimes V_2).$$

For $x \in V_1$ and $y \in V_2$, we then have

$$\begin{aligned}
\langle \langle F \rangle_\otimes, \text{proj}\,[x \otimes y] \rangle &= \text{trace}\,(\,\text{proj}\,[x \otimes y] \langle F \rangle_\otimes) \\
&= \langle (A^\top A - I)(x \otimes y), x \otimes y \rangle \\
&= |A(x \otimes y)|^2 - |x \otimes y|^2 \\
&= |F(x,y)|^2 - |x|^2 |y|^2 = 0
\end{aligned}$$

so that $\langle F \rangle_\otimes \in \mathcal{E}_{Is_{V_1} \otimes Is_{V_2}}$. That the parametrization is a bijection on the set of orthogonal equivalence classes can be shown by the argument given in the proof of Theorem 2.1. $\checkmark$

46

REMARK: Since $I_{S_{V_1}} \otimes I_{S_{V_2}} : S_{V_1} \times S_{V_2} \to S_{V_1 \otimes V_2}$ is equivariant with respect to the homomorphism $\otimes : O(V_1) \times O(V_2) \to O(V_1 \otimes V_2)$, on $\mathcal{E}_{I_{S_{V_1}} \otimes I_{S_{V_2}}} = so(V_1) \otimes so(V_2)$, the $O(V_1) \times O(V_2)$-module structure is given by $Ad \otimes Ad$.

EXAMPLE 6.4: Setting $V_1 = V_2 = \mathbf{R}^2$, by Theorem 6.1, we have

$$\mathcal{E}_{I_{S^1} \otimes I_{S^1}} = so(2) \otimes so(2) \cong \mathbf{R}$$

and the isomorphism is given by associating to $a \in \mathbf{R}$ the matrix

(6.9)
$$\begin{bmatrix} 0 & 0 & 0 & a \\ 0 & 0 & -a & 0 \\ 0 & -a & 0 & 0 \\ a & 0 & 0 & 0 \end{bmatrix}$$

This matrix is in $\mathcal{L}_{I_{S^1} \otimes I_{S^1}}$ iff $|a| \le 1$. We note that $a = \pm 1$ correspond to the orthogonal multiplications $(z, w) \to z\bar{w}$ and $(z, w) \to zw$, $z, w \in \mathbf{C} = \mathbf{R}^2$. Indeed, $z \otimes w = (xu, yu, xv, yv)$, where $z = x + iy$ and $w = u + iv$ so that we have

$$A = \begin{bmatrix} 1 & 0 & 0 & \pm 1 \\ 0 & 1 & \mp 1 & 0 \end{bmatrix}$$

Working out $A^\top A - I_4$ we obtain (6.9) for $a = \pm 1$.

EXAMPLE 6.5: Setting $V_1 = \mathbf{R}^2$ and $V_2 = \mathbf{R}^3$, by Theorem 6.1, we have

$$\mathcal{E}_{I_{S^1} \otimes I_{S^2}} = so(2) \otimes so(3) \cong \mathbf{R}^3$$

and the isomorphism is given by associating to $(a, b, c) \in \mathbf{R}^3$ the matrix

$$\begin{bmatrix} 0 & 0 & 0 & a & 0 & b \\ 0 & 0 & -a & 0 & -b & 0 \\ 0 & -a & 0 & 0 & 0 & c \\ a & 0 & 0 & 0 & -c & 0 \\ 0 & -b & 0 & -c & 0 & 0 \\ b & 0 & c & 0 & 0 & 0 \end{bmatrix}$$

Little inspection shows that this matrix is in $\mathcal{L}_{I_{S^1} \otimes I_{S^2}}$ iff $a^2 + b^2 + c^2 \leq 1$. Geometrically speaking, the moduli space $\mathcal{L}_{I_{S^1} \otimes I_{S^2}} \subset so(2) \otimes so(3)$ is the solid ball around the origin of radius 2 (with respect to the scalar product (2.1) on $S^2(\mathbf{R}^6)$).

PROBLEM: Determine the orthogonal multiplications that correspond to the points on the boundary of the moduli space $\mathcal{L}_{I_{S^1} \otimes I_{S^2}}$. (Hint: Consider quaternionic multiplication.)

REMARK: Higher dimensional examples are more subtle. For $V_1 = V_2 = \mathbf{R}^3$ the computation is still elementary (cf. [Parker]).

Orthogonal multiplications can be used to manufacture λ_2-eigenmaps between spheres in the same way as the Hopf map $H : S^3 \rightarrow S^2$ can be derived from the orthogonal multiplication $(z, w) \rightarrow z\bar{w}$, $z, w \in \mathbf{C}$ (cf. (4.3)). The general procedure that we wish to describe now is known as the *Hopf-Whitehead construction*. Let V and W be Euclidean vector spaces and

$$F : V \times V \rightarrow W$$

a full orthogonal multiplication. We define

$$f_F : S_{V \times V} \rightarrow \mathbf{R} \times W$$

by

$$f_F(x, y) = (|x|^2 - |y|^2, 2F(x, y)), \quad x, y \in V, \ |x|^2 + |y|^2 = 1.$$

Using (6.7), we have

$$|f_F(x, y)|^2 = (|x|^2 - |y|^2)^2 + 4|F(x, y)|^2$$
$$= (|x|^2 - |y|^2)^2 + 4|x|^2|y|^2 = (|x|^2 + |y|^2)^2 = 1$$

so that, restricting the image, we obtain the full λ_2-eigenmap

$$f_F : S_{V \times V} \rightarrow S_{\mathbf{R} \times W}.$$

48

In fact, the first component of f_F is clearly a harmonic quadratic polynomial on $V \times V$. The same is true for the components of F, with respect to any basis in W, since F is bilinear.

We say that f_F is the λ_2-eigenmap obtained from F by the Hopf-Whitehead construction.

PROPOSITION 6.6. *There is a linear isomorphism*

$$\Phi : \mathcal{E}_{I_{S_V} \otimes I_{S_V}} \to \mathcal{E}_{f_\otimes}$$

under which $\mathcal{L}_{I_{S_V} \otimes I_{S_V}}$ *corresponds to* $\mathcal{L}_{f_\otimes}$.

PROOF: Let

$$\Phi : S^2(V \otimes V) \to S^2(\mathbf{R} \times (V \otimes V))$$

be given by

$$\Phi(C) = [0] \times C, \; C \in S^2(V \otimes V).$$

Easy computation shows that Φ restricts to an injective linear map

$$\Phi : \mathcal{E}_{I_{S_V} \otimes I_{S_V}} \to \mathcal{E}_{f_\otimes}.$$

To prove surjectivity, let $\tilde{C} \in \mathcal{E}_{f_\otimes}$, i.e.

(6.10)
$$\langle \tilde{C} f_\otimes, f_\otimes \rangle = 0.$$

We decompose $\tilde{C}$ as

$$\begin{bmatrix} a & b^\mathsf{T} \\ b & C \end{bmatrix},$$

where $a \in \mathbf{R}$, $b \in V \otimes V$ and $C \in S^2(V \otimes V)$. Using this, (6.10) then becomes

(6.11) $a(|x|^2 - |y|^2)^2 + 4\langle b, x \otimes y \rangle(|x|^2 - |y|^2) + 4\langle C(x \otimes y), x \otimes y \rangle = 0.$

This holds on $V \times V$ as each term is a quadratic polynomial. By homogeneity in x and y, $a = 0$ and we have $\langle b, x \otimes y \rangle = 0$ so that $b = 0$. Substituting these into (6.11), we obtain that $\tilde{C} = [0] \times C$ with $C \in \mathcal{E}_{I_{S_V} \otimes I_{S_V}}$ and the first statement follows. The second is then clear from the definition of Φ. $\sqrt{}$

EXAMPLE 6.7: For $V = \mathbf{R}^2$, applying the Hopf-Whitehead construction to the tensor product

$$\otimes : \mathbf{R}^2 \times \mathbf{R}^2 \to \mathbf{R}^4$$

gives rise to the λ_2-eigenmap

$$f_\otimes : S^3 \to S^4$$

given, in real coordinates $(x, y, u, v) \in S^3 \subset \mathbf{R}^4$, by

$$(6.12) \qquad f_\otimes = (x^2 + y^2 - u^2 - v^2, 2xu, 2xv, 2yu, 2yv).$$

By Example 6.4 and Proposition 6.6, the moduli space $\mathcal{L}_{f_\otimes}$ is a segment whose vertices correspond to the Hopf map $H : S^3 \to S^2$ given in (4.3) and its 'dual' $H' : S^3 \to S^2$ given, in complex coordinates $z = x + iy$ and $w = u + iv$, by

$$(6.13) \qquad H'(z, w) = (|z|^2 - |w|^2, 2zw), \ z, w \in \mathbf{C}, \ |z|^2 + |w|^2 = 1.$$

In terms of the moduli space $\mathcal{L}_{f_{\lambda_2}}$ discussed in the previous section, $\mathcal{L}_{f_\otimes}$ corresponds, by Proposition 2.5, to the cell $I_{f_\otimes}$ that is a closed segment connecting $\langle H \rangle_{f_{\lambda_2}}$ and $\langle H' \rangle_{f_{\lambda_2}}$. These vertices lie on different components of the orbit $O(4) \cdot \langle H \rangle_{f_{\lambda_2}} = \mathbf{R}P_1^2 \cup \mathbf{R}P_2^2$ since conjugation is orientation reversing.

PROBLEM: Determine the symmetry group $SO(4)_{f_\otimes}$.

Now let $f_l : M_l \to S_{V_l}$, $l = 1, 2$, be full minimal immersions. Easy computation then shows that the tensor product $f_1 \otimes f_2 : M_1 \times M_2 \to S_{V_1 \otimes V_2}$ is also a full minimal immersion.

PROPOSITION 6.8. We have

$$\mathcal{F}_{I_{S_{V_1}} \otimes I_{S_{V_2}}} = 0.$$

50

PROOF: By Theorems 2.7 and 6.1

$$\mathcal{F}_{I_{S_{V_1}} \otimes I_{S_{V_2}}} \subset \mathcal{E}_{I_{S_{V_1}} \otimes I_{S_{V_2}}} = so(V_1) \otimes so(V_2)$$

so that an element $C \in \mathcal{F}_{I_{S_{V_1}} \otimes I_{S_{V_2}}}$ can be written as

$$C = \sum_{i=1}^{r} B_1^i \otimes B_2^i,$$

where $B_l^i \in so(V_l)$, $l = 1, 2$. Without loss of generality we may assume that $\{B_2^i\}_{i=1}^{r}$ is a linearly independent set. Since $C \in \mathcal{F}_{I_{S_{V_1}} \otimes I_{S_{V_2}}}$, for $X_x \in T_x(S_{V_1})$ and $Y_y \in T_y(S_{V_2})$, we have

(6.14) $$\sum_{i=1}^{r} \langle (B_1^i \otimes B_2^i)(\check{X}_x \otimes y + x \otimes \check{Y}_y), \check{X}_x \otimes y + x \otimes \check{Y}_y \rangle = 0.$$

This is because

$$(I_{S_{V_1}} \otimes I_{S_{V_2}})_*(X_x, Y_y)\check{} = \check{X}_x \otimes y + x \otimes \check{Y}_y.$$

By skew-symmetry, (6.14) reduces to

$$\sum_{i=1}^{r} \langle B_1^i x, \check{X}_x \rangle \langle B_2^i y, \check{Y}_y \rangle = 0.$$

Again by skew-symmetry, $\check{Y}_y$ can be *any* vector in V_2 so that we obtain

$$\sum_{i=1}^{r} \langle B_1^i x, \check{X}_x \rangle B_2^i y = 0.$$

Now y is arbitrary and hence, by linear independence, we get

$$\langle B_1^i x, \check{X}_x \rangle = 0, \; i = 1, \ldots, r.$$

Here again $\check{X}_x$ can be any vector in V_1 so that finally we obtain that $B_1^i = 0$ for all $i = 1, \ldots, r$ and $C = 0$ follows. $\checkmark$

51

§7. Direct sum

Let M be a Riemannian manifold and λ an eigenvalue of the Laplace-Beltrami operator Δ^M acting on functions on M. Let $f_1 : M \to S_{V_1}$ and $f_2 : M \to S_{V_2}$ be full λ-eigenmaps, where V_1 and V_2 are Euclidean vector spaces. We define the *direct sum*

(7.1) $$f_1 \oplus f_2 : M \to S_V$$

of f_1 and f_2 by

(7.2) $$f_1 \oplus f_2 = (f_1 \times f_2) \circ \operatorname{diag},$$

where $\operatorname{diag} : M \to M \times M$, $\operatorname{diag}(x) = (x, x)$, $x \in M$, is the diagonal map and the image of the right hand side of (7.2) is restricted to

$$V = \operatorname{span}\{(f_1 \times f_2)(x, x) | x \in M\} \subset V_1 \times V_2$$

to make (7.1) full. The direct sum $f_1 \oplus f_2 : M \to S_V$ is clearly a λ-eigenmap satisfying $f_1, f_2 \leftharpoonup f_1 \oplus f_2$.

The definition clearly extends to the direct sum of finitely many eigenmaps.

Comparing the direct sum and the direct product we arrive easily at the following :

PROPOSITION 7.1. *Let $f_1 : M \to S_{V_1}$ and $f_2 : M \to S_{V_2}$ be full λ-eigenmaps and assume that for their direct sum $f_1 \oplus f_2 : M \to S_V$ we have $V = V_1 \times V_2$. Then we have*

(7.3) $$\mathcal{E}_{f_1 \times f_2} \subset \mathcal{E}_{f_1 \oplus f_2}$$

so that

$$\mathcal{L}_{f_1 \times f_2} = \mathcal{E}_{f_1 \times f_2} \cap \mathcal{L}_{f_1 \oplus f_2}.\checkmark$$

PROBLEM: Let $f_1 : M \to S_{V_1}$ and $f_2 : M \to S_{V_2}$ be full λ-eigenmaps with direct sum $f_1 \oplus f_2 : M \to S_V$, $V \subset V_1 \times V_2$. Denoting by $P : V_1 \times V_2 \to V$ the orthogonal projection, show that the linear map $\Xi : S^2(V_1 \times V_2) \to S^2 V$, $\Xi(C) = P \cdot C \cdot P^{\mathsf{T}}$, $C \in S^2(V_1 \times V_2)$, restricts to a linear map $\Xi : \mathcal{E}_{f_1 \times f_2} \to \mathcal{E}_{f_1 \oplus f_2}$. Determine the kernel of Ξ. (Hint: $C \in \ker \Xi$ iff $C = (C_1 + cI_{V_1}) \times (C_2 - cI_{V_2})$, $C_l \in \mathcal{E}_{f_l}$, $c \in \mathbf{R}$, and

$$A_1^{\mathsf{T}}(C_1 + cI_{V_1})A_1 + A_2^{\mathsf{T}}(C_2 - cI_{V_2})A_2 = 0,$$

where $f_l = A_l(f_1 \oplus f_2)$, $l = 1, 2$.)

REMARKS: 1. The hypothesis in Proposition 7.1 is equivalent to the requirement that the components of f_1 and f_2 with respect to some bases in V_1 and V_2 form a linearly independent set of eigenfunctions on M corresponding to the eigenvalue λ.

2. The hypothesis in Proposition 7.1 is essential. In fact, if $f_1 = f_2 = I_{S_V}$, for a Euclidean vector space V then, $I_{S_V} \oplus I_{S_V}$ being orthogonally equivalent to I_{S_V}, we have $\mathcal{E}_{I_{S_V} \oplus I_{S_V}} = \{0\}$. On the other hand, by Example 5.3, $\mathcal{E}_{I_{S_V} \times I_{S_V}} \cong \mathbf{R}$.

3. The inclusion in (7.3) is, in general, proper. As an example, consider the λ-eigenmaps $H_\alpha : S^3 \to S^2$ introduced in (4.13). The direct sum $H_\alpha \oplus H_{\alpha+\pi/2}$ is orthogonally equivalent to the complex Veronese map $V : S^3 \to S^5$ given in (4.12) so that $\mathcal{E}_{H_\alpha \oplus H_{\alpha+\pi/2}} = \mathcal{E}_V \cong \mathbf{R}^2$. On the other hand, by Theorem 5.1, we have

$$\mathcal{E}_{H_\alpha \times H_{\alpha+\pi/2}} \cong \mathcal{E}_{H_\alpha} \times \mathcal{E}_{H_{\alpha+\pi/2}} \times \mathbf{R} \cong \mathcal{E}_H \times \mathcal{E}_H \times \mathbf{R} \cong \mathbf{R}.$$

The geometry behind this is clear : $\mathcal{L}_V$ is a 2-disk with center the origin $0 = \langle V \rangle_V$ and $\langle H_\alpha \rangle_V$ and $\langle H_{\alpha+\pi/2} \rangle_V$ are antipodal points on the boundary circle $\partial \mathcal{L}_V$. Their span is a diagonal segment in the disk.

EXAMPLE 7.2: The direct sum of the Hopf map $H : S^3 \to S^2$ and its dual $H' : S^3 \to S^2$ is orthogonally equivalent with the eigenmap $f_\otimes : S^3 \to S^4$ associated to the tensor product $\otimes : \mathbf{R}^2 \times \mathbf{R}^2 \to \mathbf{R}^4$ by the Hopf-Whitehead construction (cf. Example 6.7).

EXAMPLE 7.3: Consider

$$f = H' \oplus V.$$

In complex coordinates $z = x + iy$ and $w = u + iv$, we have

$$f(z, w) = (\frac{1}{\sqrt{2}}(|z|^2 - |w|^2), \frac{1}{\sqrt{2}}z^2, \frac{1}{\sqrt{2}}w^2, \sqrt{3}zw)$$

so that

$$f : S^3 \to S^6$$

is a full λ_2-eigenmap. Computation shows that $C \in \mathcal{E}_f \subset S^2(\mathbf{R}^7)$ iff $C = [0] \times C(a, b, c)$, $a, b, c \in \mathbf{R}$, where $C(a, b, c)$ is given by

$$\begin{bmatrix} -a & b & 0 & 0 & 0 & c \\ b & -a & 0 & 0 & -c & 0 \\ 0 & 0 & -a & 0 & 0 & 0 \\ 0 & 0 & 0 & -a & 0 & 0 \\ 0 & -c & 0 & 0 & (a-b)/3 & c/3 \\ c & 0 & 0 & 0 & c/3 & (a+b)/3 \end{bmatrix}$$

In particular $\mathcal{E}_f \cong \mathbf{R}^3$. Using (a, b, c) as coordinates on $\mathbf{R}^3$ and evaluating $C + I_6 \geq 0$, we obtain that $\mathcal{L}_f \subset \mathbf{R}^3$ is a finite cone with vertex $\langle H' \rangle_f = (1, 0, 0)$ and base circle $\{\langle H_\alpha \rangle_f | \alpha \in \mathbf{R}\}$ of radius 2 and center $\langle V \rangle_f = (-1, 0, 0)$ which is perpendicular to the a-axis.

54

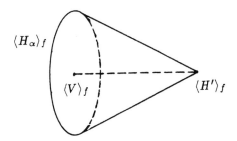

$\langle H_\alpha \rangle_f$ $\langle V \rangle_f$ $\langle H' \rangle_f$

The center $S^1 \subset U(2)$ rotates the cone. The interior, as usual, corresponds to all full λ_2-eigenmaps $f' : S^3 \to S^6$ with $f' \hookleftarrow f$. Finally, the boundary lines correspond to those full λ_2-eigenmaps $f'' : S^3 \to S^4$ that are geometrically not distinct from $f_\otimes : S^3 \to S^4$.

PROBLEM: Derive an explicit expression of the direct sum of the Hopf map and the Veronese map.

EXAMPLE 7.4: The 'dual' of the complex Veronese map V is the full λ_2-eigenmap $V' : S^3 \to S^5$ defined by

$$V'(z, w) = (z^2, \sqrt{2}z\bar{w}, w^2).$$

As for V, the moduli space $\mathcal{L}_{V'} \subset \mathcal{E}_{V'}$ is a 2-dimensional disk with boundary $\partial \mathcal{L}_{V'} = \{\langle H'_\alpha \rangle_{V'} | \alpha \in \mathbf{R}\}$, where

$$H'_\alpha = S^3 \to S^2$$

is given by

$$H'_\alpha = (e^{2i\alpha}z^2 + w^2, \operatorname{Im}(2e^{i\alpha}z\bar{w})).$$

We have

$$\operatorname{diag}(e^{i\theta}, e^{i\theta})\langle H'_\alpha \rangle_{V'} = \langle H'_\alpha \rangle_{V'}$$

so that the diagonal center $S^1 \subset U(2)$, while rotating $\mathcal{L}_V$ (cf.(4.14)), leaves $\mathcal{L}_{V'}$ pointwise fixed. Similarly, the *antidiagonal* $\{\,\mathrm{diag}\,(e^{i\theta}, e^{-i\theta})|\theta \in \mathbf{R}\}$ rotates $\mathcal{L}_{V'}$ and leaves $\mathcal{L}_V$ pointwise fixed.

The direct sum of V and V' is a full λ_2-eigenmap

$$V \oplus V' : S^3 \to S^7$$

given, in real coordinates $(x, y, u, v) \in S^3 \subset \mathbf{R}^4$, by

$$V \oplus V'(x, y, u, v) = (x^2 - y^2, 2xy, \sqrt{2}xu, \sqrt{2}xv, \sqrt{2}yu, \sqrt{2}yv, u^2 - v^2, 2uv).$$

Computation shows that $\bar{I}_{V \oplus V'}$ is 5-dimensional and is nothing but the convex hull of the 2-disks $\bar{I}_V$ and $\bar{I}_{V'}$ in $\mathcal{L}_{f_{\lambda_2}} \subset \mathcal{E}_{f_{\lambda_2}}$.

CHAPTER II

EIGENMAPS AND MINIMAL IMMERSIONS OF A RIEMANNIAN HOMOGENEOUS SPACE INTO SPHERES

§1. First things first on homogeneous spaces

We begin with a *compact homogeneous space* M that is a compact manifold M on which a compact Lie group G acts transitively. Given a *base point* $o \in M$ let $K \subset G$ denote the *isotropy subgroup* of the action of G on M at o. Equivalently, K consists of those elements of G that leave o fixed. Taking differentials at o we obtain that the tangent space $T_o(M)$ is a K-module. M is said to be *isotropy irreducible* if $T_o(M)$ is an irreducible K-module.

In general, we can (and will) identify M with the quotient G/K by the C^∞-diffeomorphism that associates to $a \cdot o$ the left-class aK, $a \in G$. The base point o then gets identified with the subgroup K. Via the identification $M = G/K$, the action of G on M corresponds to the action of G on G/K by left translations. We will use the symbols M and G/K alternately in the sequel. Note that G/K comes equipped with the natural projection $\pi : G \to G/K$ that associates to $a \in G$ the left-class aK. Clearly, π is equivariant (with respect to the actions of G on itself and on G/K by left translations) and hence it is a C^∞-submersion, i.e. a C^∞-map of maximal rank everywhere.

Let $\mathcal{G}$ denote the Lie algebra of G. We think of $\mathcal{G}$ as being the Lie algebra of *right-invariant* vector fields on G or, by restriction, the tangent space of G at the identity

57

element 1. The primary reason for taking right-invariant vector fields on G is that their induced 1-parameter groups of transformations consist of left translations.

The Lie algebra $\mathcal{K}$ of the closed subgroup $K \subset G$ is a Lie subalgebra of $\mathcal{G}$. *Any* linear subspace $\mathcal{P} \subset \mathcal{G}$ that is complementary to $\mathcal{K} \subset \mathcal{G}$ can be identified with the tangent space $T_o(G/K)$. The identification is given by restricting π_* to $\mathcal{P}$ that establishes a linear isomorphism between $\mathcal{P}$ and $T_o(G/K)$.

The Lie algebra $\mathcal{G}$ carries a natural G-module structure by the *adjoint representation Ad* (that is the differential (at 1) of the action of G on itself by inner automorphisms). Let $\mathcal{P} \subset \mathcal{G}$ be a K-submodule that is complementary to $\mathcal{K}$ in $\mathcal{G}$. Equivalently, let $\mathcal{G} = \mathcal{K} \oplus \mathcal{P}$ be a decomposition of $\mathcal{G}$ into K-submodules, where the K-module structure on $\mathcal{G}$ is given by the restriction $Ad|_K$. Then the trivial observation that K stays invariant by left translations by elements of K translates into the fact that the restriction $\pi_*|_{\mathcal{P}} : \mathcal{P} \to T_o(G/K)$ is a K-module isomorphism.

Let $M = G/K$ be a compact homogeneous space. A Riemannian metric g on M is said to be *Riemannian homogeneous* if G acts on M by isometries. g is then uniquely determined by its restriction to $T_o(M)$ (or on any K-submodule $\mathcal{P} \subset \mathcal{G}$ complementary to $\mathcal{K}$) on which it is a K-invariant scalar product that makes $T_o(M)$ an orthogonal K-module. If M is isotropy irreducible then g is unique up to a positive constant multiple according to the following:

PROPOSITION 1.1. *Let L be a compact Lie group and W an irreducible L-module. Then the linear space of L-invariant symmetric bilinear forms is 1-dimensional.* $\sqrt{}$

We now turn the setting around and begin with a compact Lie group G and a scalar product on its Lie algebra $\mathcal{G}$ that is assumed to be invariant under the adjoint representation of G on $\mathcal{G}$. Given a closed subgroup $K \subset G$ with Lie algebra $\mathcal{K} \subset \mathcal{G}$, we define $\mathcal{P} = \mathcal{K}^\perp \subset \mathcal{G}$. By invariance, $\mathcal{P}$ is a K-submodule of $\mathcal{G}$ and the previous construction applies, i.e. the restriction of the scalar product to $\mathcal{P}$ makes $M = G/K$

a Riemannian homogeneous space. The Riemannian homogeneous metric (or space) obtained in this way is said to be *naturally reductive*. Note that in the special case when $K = \{1\}$, a naturally reductive Riemannian metric on G is nothing but a biinvariant Riemannian metric on G, i.e. a Riemannian metric on G that is invariant under both left and right translations. By equivariance, the natural projection $\pi : G \to G/K$ becomes a Riemannian submersion with respect to the naturally reductive Riemannian metrics on G and on G/K. In general, a C^∞-submersion $f : \bar{M} \to M$ between Riemannian manifolds is said to be a *Riemannian submersion* if the differential of f restricted to the *horizontal distribution* $H = (\ker f_*)^\perp \subset T(\bar{M})$ is isometric.

PROPOSITION 1.2. *A Riemannian submersion $f : \bar{M} \to M$ is harmonic iff the fibres of f are minimal submanifolds of $\bar{M}$. In this case we have*

$$(1.1) \qquad \Delta^{\bar{M}}(\mu \circ f) = (\Delta^M \mu) \circ f$$

for any C^∞-function μ on M.

PROOF: Both statements are easy consequences of the general composition formula (that we managed to avoid up to this point). In fact, for any C^∞-maps $h : N \to N'$ and $h' : N' \to N''$ between Riemannian manifolds, the second fundamental forms relate as

$$(1.2) \qquad \beta(h' \circ h) = h'_* \circ \beta(h) + \beta(h')(h_*, h_*).$$

The proof of (1.2) is a challenging local computation (cf. [Eells-Sampson]). Taking traces, we obtain

$$(1.3) \qquad \tau(h' \circ h) = h'_* \circ \tau(h) + \operatorname{trace} \beta(h')(h_*, h_*).$$

Turning to the proof of the proposition, let F be a fibre of f with inclusion map $\iota : F \to \bar{M}$. Applying (1.3), we obtain

$$\tau(f \circ \iota) = f_* \circ \tau(\iota) + \operatorname{trace} \beta(f)(\iota_*, \iota_*).$$

The left hand side is zero since $f \circ \iota$ is constant. Moreover, $\tau(\iota)$ is a section of the horizontal distribution H of $f : \bar{M} \to M$ since ι is isometric. We obtain that ι is minimal iff

$$(1.4) \qquad\qquad \text{trace}\, \beta(f)(\iota_*, \iota_*) = 0.$$

On the other hand, using e.g. orthonormal moving frames, it follows by easy computation that the trace of $\beta(f)$ on H is always zero. Hence (1.4) holds iff trace $\beta(f) = 0$ iff f is harmonic. The first statement follows.

Assume now that $f : \bar{M} \to M$ is harmonic and let μ be a C^∞-function on M. By (1.3), we have

$$\Delta^{\bar{M}}(\mu \circ f) = -\,\text{trace}\,\beta(\mu \circ f) = -\tau(\mu \circ f)$$

$$= -\mu_* \circ \tau(f) - \text{trace}\,\beta(\mu)(f_*, f_*)$$

$$= -\,\text{trace}\,\beta(\mu) \circ f = (\Delta^M \mu) \circ f$$

and (1.1) follows. $\checkmark$

REMARK: The composition formula (1.3) reveals many basic (composition) properties of harmonic maps. For the time being we mention here only the following:

1. The composition of harmonic maps is, in general, not harmonic. Actually, we know this from §4 of Chapter I. A periodic geodesic with rotation number $\neq \pm 1$ on a *flat* Clifford torus is not a geodesic in S^3.

2. The composition of a harmonic map with a *totally geodesic map* (i.e. a map whose second fundamental form vanishes) is a harmonic map.

3. The composition formula (1.3) also implies that if $h : N \to N'$ is a harmonic Riemannian submersion then $h' : N' \to N''$ is harmonic iff $h' \circ h$ is harmonic.

(ANOTHER) REMARK: Let $f : \bar{M} \to M$ be a Riemannian submersion with totally geodesic fibres. A result of [Hermann] states that the horizontal lift of a geodesic

60

on M (with respect to H) is a geodesic on $\bar{M}$. Using this and working out the Laplace-Beltrami operator in terms of geodesics, (1.1) follows easily (cf. [Wallach]).

Proposition 1.2 applies to the natural projection $\pi : G \to G/K$ since, with respect to the naturally reductive Riemannian metrics, it is a Riemannian submersion and the fibres, being just the left-classes of K, are totally geodesic and hence minimal. (Actually, the geodesics of G are nothing but the left and right translates of the 1-parameter subgroups.) We obtain that π is a harmonic map and, for any C^∞-function μ on G/K, we have

$$(1.5) \qquad \Delta^G(\mu \circ \pi) = (\Delta^{G/K}\mu) \circ \pi.$$

We now wish to describe the Laplace-Beltrami operators Δ^G and $\Delta^{G/K}$ in terms of the Lie algebra $\mathcal{G}$. Recall that $\mathcal{G}$ is the Lie algebra of right-invariant vector fields on G so that any vector in $\mathcal{G}$ can be thought of as a first order differential operator acting on C^∞-functions on G.

PROPOSITION 1.3. *With respect to a biinvariant Riemannian metric on G, we have*

$$(1.6) \qquad \Delta^G = -\sum_{i=1}^{m} E_i^2,$$

where $\{E_i\}_{i=1}^{m} \subset \mathcal{G}$ is an orthonormal basis.

PROOF: First observe that the Levi-Civita connection ∇^G of the biinvariant Riemannian metric on G can be defined by

$$(1.7) \qquad \nabla_X^G Y = \frac{1}{2}[X,Y], \ X, Y \in \mathcal{G}.$$

(In fact, by unicity, we only have to check that ∇^G given in (1.7) is torsionfree and preserves the metric.) Thus, for a C^∞-function $\bar{\mu}$ on G, we have

$$\Delta^G\bar{\mu} = -\sum_{i=1}^{m}(\nabla_{E_i}^G\bar{\mu}_*)E_i = -\sum_{i=1}^{m}\nabla_{E_i}^G(\bar{\mu}_*E_i) + \sum_{i=1}^{m}\bar{\mu}_*(\nabla_{E_i}^G E_i) = -\sum_{i=1}^{m}E_i^2\bar{\mu},$$

since $\nabla^G_{E_i} E_i = \frac{1}{2}[E_i, E_i] = 0$ by skew-symmetry. $\sqrt{}$

We now wish to derive a formula similar to (1.6) for $\triangle^{G/K}$. Let $X \in \mathcal{G}$ be a right-invariant vector field on G. As a first order differential operator, X acts on a C^∞-function $\bar{\mu}$ on G by the formula

$$X \cdot \bar{\mu} = \frac{d}{dt}(\bar{\mu} \circ \exp^G(tX))|_{t=0}$$

where $\exp^G : \mathcal{G} \to G$ denotes the exponential map of the Lie group G. This is because the 1-parameter group of transformations induced by X consists of the left translations by $\exp^G(tX)$, $t \in \mathbf{R}$.

Keeping in mind the equivariance of the natural projection $\pi : G \to G/K$, for $X \in \mathcal{G}$, we define X^* to be the vector field on G/K that acts as a first order differential operator on a C^∞-function μ on G/K by

(1.8) $$X^* \cdot \mu = \frac{d}{dt}(\mu \circ \exp^G(tX))|_{t=0}.$$

Actually, X^* is the unique vector field on G/K such that X and X^* are π-related, i.e. $\pi_* X = X^* \circ \pi$.

PROPOSITION 1.4. *With respect to a naturally reductive Riemannian matric on G/K, we have*

(1.9) $$\triangle^{G/K} = -\sum_{i=1}^{m}(E_i^*)^2,$$

where $\{E_i\}_{i=1}^{m} \subset \mathcal{G}$ is an orthonormal basis.

PROOF: Let μ be a C^∞-function on G/K. Combining (1.5) and (1.6) we have

$$(\triangle^{G/K}\mu) \circ \pi = \triangle^G(\mu \circ \pi) = -\sum_{i=1}^{m} E_i^2(\mu \circ \pi) = -\sum_{i=1}^{m} \frac{d^2}{dt^2}(\mu \circ \pi \circ \exp^G(tE_i))|_{t=0}$$

$$= -\sum_{i=1}^{m} \frac{d^2}{dt^2}(\mu \circ \exp^G(tE_i) \circ \pi)|_{t=0} = -\sum_{i=1}^{m}(E_i^*)^2 \cdot \mu \circ \pi$$

62

and (1.9) follows. $\checkmark$

§2. Equivariant λ-eigenmaps and minimal immersions

Let $f : M \to S_V$ be a C^∞-map of a compact homogeneous space $M = G/K$ into the unit sphere of a Euclidean vector space V. Then f is said to be *equivariant* if there exists a homomorphism $\rho : G \to O(V)$ such that f is equivariant with respect to ρ, i.e. we have

$$(2.1) \qquad f(a \cdot o) = \rho(a) \cdot v^0, \, a \in G,$$

where $o \in M$ is the base point corresponding to K and $v^0 = f(o) \in S_V$. Clearly ρ and the unit vector v^0 determine f uniquely. The homomorphism ρ makes V an orthogonal G-module. f is full iff v^0 is not contained in any proper G-submodule of V. This is certainly the case when V is irreducible. Note that if f is full then ρ is uniquely determined by f.

EXAMPLE 2.1: The role of G is crucial. For example, the Hopf map $H : S^3 \to S^2$ discussed in §4 of Chapter I is equivariant for $G = U(2)$ but not equivariant for $G = SO(4)$. This observation will play an important part in the sequel.

PROBLEM*: Let $M = G/K$ be a compact naturally reductive Riemannian homogeneous space and $f : M \to S_V$ a full harmonic imbedding that is equivariant with respect to a monomorphism $\rho : G \to O(V)$. Let v be a vector field along f and assume that v is G-invariant, i.e. $v_{a \cdot x} = \rho(a)_* \cdot v_x$, $x \in M$, $a \in G$. Show that v is a divergencefree Jacobi field along f, in particular, v is a harmonic variation along f, i.e. $f_t = \exp \circ (tv) : M \to S_V$ is harmonic for all $t \in \mathbf{R}$ (cf. Problem* after Proposition 1.2 of Chapter I.)(Hint: Show that $\Delta^G \rho = e(f) \cdot \rho$ with respect to the biinvariant Riemannian metric on G.)

We now reverse the procedure and begin with an orthogonal G-module V and a unit vector $v^0 \in V$ which we assume to be left fixed by a closed subgroup $K \subset G$. We denote by $\rho : G \to O(V)$ the homomorphism that defines the G-module structure on V. Setting $M = G/K$, we *define* the C^∞-map $f_V : G/K \to S_V$ by (2.1) (replacing f by f_V). Since the right hand side of (2.1) depends only on the left-class aK, $a \in K$, the map f_V is well-defined. By birth, f_V is equivariant with respect to ρ. Putting a naturally reductive Riemannian metric on G/K we wish to study when is $f_V : G/K \to S_V$ a λ-eigenmap for some $\lambda \in \mathbf{R}$. To do this we need a preliminary construction.

Let $C^\infty(M)$ denote the linear space of all C^∞-functions on M. Then $C^\infty(G/K)$ is a G-module where the action of $a \in G$ on $\mu \in C^\infty$ is given by

(2.2)
$$a \cdot \mu = \mu \circ a^{-1}.$$

We define the linear map

$$\Phi_V : V \to C^\infty(G/K)$$

by

(2.3)
$$\Phi_V(v) = \langle f_V, v \rangle.$$

We claim that Φ_V is a homomorphism of G-modules. Indeed, for $a \in G$, we have

$$\Phi_V(\rho(a)v) = \langle f_V, \rho(a)v \rangle = \langle \rho(a)^{-1} f_V, v \rangle$$
$$= \langle f_V \circ a^{-1}, v \rangle = \langle f_V, v \rangle \circ a^{-1}$$
$$= a \cdot \langle f_V, v \rangle.$$

If V is an irreducible G-module then Φ_V is an isomorphism onto its image in $C^\infty(G/K)$. This latter fact deserves a special attention so that we introduce the following terminology: Given a compact Lie group G and a closed subgroup $K \subset G$,

64

an *irreducible* orthogonal G-module V is said to be *class 1 with respect to the pair* (G, K) if V contains a nonzero vector v^0 left fixed by K, or equivalently, if the restriction $V|_K$ contains the trivial K-module. We have just observed that if V is a class 1 module with respect to the pair (G, K) then Φ_V maps V isomorphically onto an irreducible component of $C^\infty(G/K)$. Conversely, any irreducible component $\tilde{V}$ of $C^\infty(G/K)$ is class 1 with respect to the pair (G, K). Indeed, let $\{f^i\}_{i=0}^n \subset \tilde{V}$ be an orthonormal basis with respect to any G-invariant scalar product on $\tilde{V}$. (For example, we can take the L^2-scalar product (restricted to $\tilde{V}$) with respect to, say, a naturally reductive Riemannian metric on G/K.) Consider $v^0 = \sum_{i=0}^n f^i(o)f^i \in \tilde{V}$. Easy computation shows that v^0 does not depend on the choice of the orthonormal basis and is therefore left fixed by K.

REMARK: Let V be a complex G-module. V is said to be *unitary* if V is endowed with a G-invariant Hermitian scalar product so that the G-module structure on V is given by a homomorphism $\rho : G \to U(V)$. An irreducible unitary G-module is said to be a complex *class 1 module with respect to the pair* (G, K) if $V|_K$ contains the trivial (complex) K-module. A theorem of É. Cartan states that if (G, K) is a symmetric pair of compact type then $\sum_{\sigma \in \Sigma} \Phi_{V_\sigma}(V_\sigma)$ is L^2-dense in $C^\infty(G/K) \otimes_\mathbf{R} \mathbf{C}$, where $\{V_\sigma\}_{\sigma \in \Sigma}$ is a complete set of mutually inequivalent complex class 1 modules with respect to the pair (G, K) (cf. [É.Cartan;1][Wallach]).

THEOREM 2.2. *Let V be a class 1 module with respect to the pair (G, K), where G is a compact Lie group and $K \subset G$ is a closed subgroup. Let $v^0 \in V$ be a unit vector left fixed by K. Then $f_V : G/K \to S_V$ given by (2.1) is a full λ-eigenmap with respect to any naturally reductive Riemannian metric on G/K. If G/K is isotropy irreducible then $f_V : G/K \to S_V$ is a full minimal immersion that induces a naturally reductive Riemannian metric on M.*

PROOF: Let $\tilde{V}$ denote the image of V under Φ_V. By the above, $\tilde{V}$ is an irreducible

65

G-submodule of $C^\infty(G/K)$ isomorphic with V. By (2.3), f_V is a λ-eigenmap iff the Laplace-Beltrami operator $\Delta^{G/K}$ acts on $\tilde{V}$ as $\lambda \cdot I$.

First we claim that $\Delta^{G/K}(\tilde{V}) \subset \tilde{V}$. Let $\mu \in \tilde{V}$. Given $X \in \mathcal{G}$, we have

$$\mu \circ \exp^G(tX) = \exp^G(-tX) \cdot \mu \in \tilde{V}, \, t \in \mathbf{R},$$

since $\tilde{V}$ is a G-submodule of $C^\infty(G/K)$. Differentiating, by (1.8), we obtain that $X^*(\mu) \in \tilde{V}$ so that X^* maps $\tilde{V}$ into itself. By (1.9), $\Delta^{G/K}$ maps $\tilde{V}$ into itself.

Secondly, we claim that the restriction of $\Delta^{G/K}$ to $\tilde{V}$ is a symmetric linear endomorphism of $\tilde{V}$ with respect to any G-invariant scalar product $\langle , \rangle$ on $\tilde{V}$. Let $\mu, \mu' \in \tilde{V}$. Given $X \in \mathcal{G}$, we have

$$\langle \mu, \mu' \rangle = \langle \exp^G(-tX) \cdot \mu, \exp^G(-tX) \cdot \mu' \rangle$$
$$= \langle \mu \circ \exp^G(tX), \mu' \circ \exp^G(tX) \rangle, \, t \in \mathbf{R}.$$

Differentiating, by (1.8), we obtain that X^* is skew on $\tilde{V}$. By (1.9), $\Delta^{G/K}$ is symmetric and hence diagonalizable on $\tilde{V}$.

Finally, $\Delta^{G/K}$ commutes with the action of G on $\tilde{V}$. Since $\tilde{V}$ is irreducible, $\Delta^{G/K}$ is a (real) constant multiple of the identity so that the first statement follows.

Assume now that G/K is isotropy irreducible and let g be a naturally reductive Riemannian metric on G/K. Consider the symmetric 2-covariant tensor field h on G/K defined by

$$h(X, Y) = \langle f_* X, f_* Y \rangle,$$

where X and Y are vector fields on G/K. By equivariance of f, the tensor field h is G-invariant. In particular, its restriction h_o to $T_o(G/K)$ is a K-invariant symmetric bilinear form. By Proposition 1.1, we have $h_o = c \cdot g_o$, where g_o is the restriction of g to $T_o(G/K)$. By G-invariance we obtain $h = c \cdot g$. We obtain that $f_V : G/K \to S_V$ is an isometric immersion of a constant multiple of the naturally reductive Riemannian metric g on G/K. (Actually, $c = \lambda/m$, $m = \dim M$, which follows by comparing the energy densities of f_V with respect to g and $c \cdot g$.) Since f_V is also harmonic (with respect to (any positive constant multiple of) g) it is minimal. $\checkmark$

As a byproduct, we obtained that any irreducible component $\tilde{V}$ of $C^\infty(G/K)$ is contained in an eigenspace V_λ corresponding to an eigenvalue λ of the Laplace-Beltrami operator of the homogeneous space $M = G/K$ with respect to *any* naturally reductive Riemannian metric on M. We now describe another method to obtain λ-eigenmaps based on V_λ rather than $\tilde{V}$.

Let $M = G/K$ be a compact Riemannian homogeneous space (not necessarily naturally reductive) and let $V_\lambda \subset C^\infty(M)$ denote the eigenspace of Δ^M corresponding to the eigenvalue λ. Then V_λ is an orthogonal G-module. The G-module structure on V_λ is given by (2.2) and we choose as a G-invariant scalar product $\langle \, , \, \rangle$ on V_λ the normalized L^2-scalar product

$$(2.4) \qquad \langle \mu, \mu' \rangle = \frac{n(\lambda) + 1}{\text{vol}(M)} \int_M \mu\mu' \cdot \nu_M, \ \mu, \mu' \in V_\lambda,$$

where $\dim V_\lambda = n(\lambda) + 1$ is the multiplicity of the eigenvalue λ. Let $\{f_\lambda^i\}_{i=0}^{n(\lambda)} \subset V_\lambda$ be an orthonormal basis, which, at the same time, identifies V_λ with $\mathbf{R}^{n(\lambda)+1}$, and define

$$f_\lambda = (f_\lambda^0, \ldots, f_\lambda^{n(\lambda)}) : M \to \mathbf{R}^{n(\lambda)+1}.$$

For each $a \in G$, $\{f_\lambda^i \circ a^{-1}\}_{i=0}^{n(\lambda)}$ is again an orthonormal basis of V_λ so that we have

$$a \cdot f_\lambda^i = f_\lambda^i \circ a^{-1} = \sum_{j=0}^{n(\lambda)} A_{ji} \cdot f_\lambda^j, \ i = 0, \ldots, n(\lambda),$$

where $A = (A_{ij})_{i,j=0}^{n(\lambda)} \in O(n(\lambda) + 1)$. Associating to a the orthogonal matrix A gives rise to a homomorphism $\rho_\lambda : G \to O(n(\lambda)+1)$ that is actually nothing but the homomorphism associated to the G-module structure of V_λ under the identification $V_\lambda \cong \mathbf{R}^{n(\lambda)+1}$. By orthogonality, we have

$$\sum_{i=0}^{n(\lambda)} (a \cdot f_\lambda^i)^2 = \sum_{i,j,k=0}^{n(\lambda)} A_{ji} A_{ki} \cdot f_\lambda^j f_\lambda^k = \sum_{i=0}^{n(\lambda)} (f_\lambda^i)^2$$

so that this function is constant on M. Integrating and using orthonormality, we obtain

$$\sum_{i=0}^{n(\lambda)} (f_\lambda^i)^2 \cdot \text{vol}(M) = \sum_{i=0}^{n(\lambda)} \int_M (f_\lambda^i)^2 \cdot \nu_M = \frac{\text{vol}(M)}{n(\lambda)+1} \sum_{i=0}^{n(\lambda)} (f_\lambda^i)^2 = \text{vol}(M)$$

so that

$$\sum_{i=0}^{n(\lambda)} (f_\lambda^i)^2 = 1.$$

In other words, f_λ defines a λ-eigenmap

$$f_\lambda : M \to S^{n(\lambda)}$$

which is equivariant with respect to the homomorphism $\rho_\lambda : G \to O(n(\lambda)+1)$. Clearly, f_λ is full and different choices of the orthonormal bases of V_λ give orthogonally equivalent full λ-eigenmaps. f_λ is said to be a *standard λ-eigenmap* associated to the eigenvalue λ of $\triangle^M$. (Cf. also [Parks-Urakawa].) If $M = G/K$ is isotropy irreducible then the second part of the proof of Theorem 2.2 applies and we obtain that $f_\lambda : M \to S^{n(\lambda)}$ is a minimal immersion of a constant multiple of the (automatically) naturally reductive Riemannian metric on M. As noted above, the value of the constant is λ/m, where $\dim M = m$.

PROBLEM *: Prove that for any compact homogeneous space $M = G/K$ there exists an equivariant minimal immersion $f : M \to S^n$ for sufficiently large n. (Hint: First use Mostow's imbedding theorem to conclude the existence of an imbedding $f : G/K \to S^n$ that is equivariant with respect to a monomorphism $\rho : G \to SO(n+1)$ (cf. [Mostow]). Denote by $\Omega \subset S^n$ the (closed) set consisting of those points whose isotropy subgroup (with respect to the action of G via ρ) contains K. Define a volume function on Ω and show that it is continuous. Maximize the volume function on Ω.)(Cf. [Wu-Yi Hsiang;1][Wallach].)

PROBLEM **: Let $G \subset O(n+1)$ be a closed subgroup and assume that a G-orbit $M \subset S^n$ is a full codimension 1 minimal submanifold in S^n. Assume that M, as an isoparametric hypersurface, has degree $p > 2$ (cf. [É.Cartan][Münzner]). Show that the Riemannian homogeneous metric on M induced from S^n is *not* naturally reductive. (This generalizes the example given by [Lawson, p.25].) (Hint: Assume that the induced Riemannian metric on M is naturally reductive. Use Problem* after Example 2.1 to show that a unit section v of the normal bundle of the imbedding $M \subset S^n$ is a harmonic variation. On the other hand, show that v is a harmonic variation iff $\sum_{i=1}^p m_i \lambda_i^2 = n - 1$, where $\lambda_i = -\cot\left(\frac{(2i-1)\pi}{2p}\right)$ denotes the i-th principal curvature of M with multiplicity m_i. Finally, use Münzner's classification theorem (cf. [Münzner]) to prove that this equation is satisfied iff $p = 2$. (Note that the possible values of p are 1,2,3,4,6 and $p = 1$ is excluded by fullness of M in S^n. Moreover $m_{i+2} = m_i$ with $m_1 + m_2 + m_1 + m_2 + \ldots = n - 1$ and $\sum_{i=1}^p m_i \lambda_i = 0$ since M is minimal in S^n.))

EXAMPLE 2.3: In §4 of Chapter I we worked out (somewhat prematurely) the standard minimal immersion $f_{\lambda_2} : S^3 \to S^8$ corresponding to the second eigenvalue λ_2 of Δ^{S^3}, where $S^3 = SO(4)/SO(3)$ is isotropy irreducible. If we think of S^3 as $U(2)/S^1$, where $S^1 \subset U(2)$ is central then V_{λ_2} is no longer irreducible as a $U(2)$-module. In fact, the components of the Hopf map $H : S^3 \to S^2$ in (4.5) of Chapter I span an irreducible $U(2)$-submodule $\tilde{V} \subset V_{\lambda_2}$ and $f_{\tilde{V}} = H$.

Returning to the general situation where $M = G/K$ is a compact Riemannian homogeneous space and $f_\lambda : M \to S^{n(\lambda)}$ is a (from here on fixed) standard λ-eigenmap we see that a full harmonic map $f : M \to S^n$ is derived from f_λ iff f is a full λ-eigenmap. This is because the components of f_λ comprise an orthonormal basis in V_λ. Hence *the standard moduli space $\mathcal{L}_\lambda = \mathcal{L}_{f_\lambda}$ parametrizes (the orthogonal equivalance classes of) all full λ-eigenmaps of M into spheres.* For $M = G/K$ isotropy irreducible, $\mathcal{M}_\lambda = \mathcal{M}_{f_\lambda}$ *parametrizes (the orthogonal equivalence classes of) all full minimal immersions of M into spheres that induce λ/m times the Riemannian metric on M.* The main objective of these notes is to study the standard

69

moduli spaces $\mathcal{L}_\lambda$ and $\mathcal{M}_\lambda$. We observe *en passant* that $so(V_\lambda)\oplus\mathcal{E}_\lambda$, where $\mathcal{E}_\lambda = \mathcal{E}_{f_\lambda}$, is isomorphic with the linear space of *all* divergencefree Jacobi fields along f_λ (cf. Proposition 2.6 of Chapter I). This is because, for a given divergencefree Jacobi field v along f_λ, the components of $\check{v}$ are in V_λ so that v is automatically derived from f_λ. Note also that $\mathcal{E}_\lambda \subset S^2(V_\lambda)$ is actually contained in the linear subspace $S_0^2(V_\lambda)$ of *traceless* symmetric endomorphisms of V_λ. This follows from (2.8) and (2.9) of Chapter I since, with respect to the isomorphism $V_\lambda \cong \mathbf{R}^{n(\lambda)+1}$, the ij-th entry of the matrix $\int_M \text{proj}\,[f_\lambda]\cdot\nu_M$ is

$$\int_M f_\lambda^i f_\lambda^j \cdot \nu_M = \frac{\text{vol}\,(M)}{n(\lambda)+1}\delta_{ij},$$

where δ_{ij} is the Kronecker symbol and we used (2.4).

§3. Class 1 modules

Let G be a compact Lie group and $K \subset G$ a closed subgroup. Given a finite dimensional K-module W we define the *induced G-module* $\text{Ind}_K^G(W)$ to be the G-module consisting of all continuous maps $w : G \to W$ that are equivariant with respect to the K-module structure of W, where K is considered to act on G by left translations. Equivalently, $w \in \text{Ind}_K^G(W)$ iff $w : G \to W$ is a continuous map satisfying

$$w(k \cdot a) = k \cdot w(a),\ k \in K, a \in G.$$

The G-module structure on $\text{Ind}_K^G(W)$ is given by $a \in G$ acting on $w \in \text{Ind}_K^G(W)$ by precomposing w with the right multiplication by a or, in other words, by setting

$$(a \cdot w)(a') = w(a' \cdot a),\ a' \in G.$$

Note that $\text{Ind}_K^G(W)$ is infinite dimensional (unless K has finite index in G).

With obvious notations, we have

$$\mathrm{Ind}_G^G(W) \cong W$$

and

$$\mathrm{Ind}_K^G(W \oplus W') \cong \mathrm{Ind}_K^G(W) \oplus \mathrm{Ind}_K^G(W').$$

Moreover

$$\mathrm{Ind}_L^G(\mathrm{Ind}_K^L(W)) \cong \mathrm{Ind}_K^G(W),$$

where $L \subset G$ is a closed subgroup that contains K.

The following elementary but crucial result is known as Frobenius Reciprocity :

PROPOSITION 3.1. *For any finite dimensional G-modules W and Z we have*

(3.1) $$\mathrm{Hom}_K(Z|_K, W) \cong \mathrm{Hom}_G(Z, \mathrm{Ind}_K^G(W)).$$

In particular, if W and Z are irreducible, we have for the multiplicities

(3.2) $$m[W : Z|_K] = m[Z : \mathrm{Ind}_K^G(W)].$$

PROOF: Let $A \in \mathrm{Hom}_K(Z|_K, W)$ and define $\tilde{A} : Z \to \mathrm{Ind}_K^G(W)$ by

$$\tilde{A}(z)(a) = A(a \cdot z), \ z \in Z, a \in G.$$

Clearly, $\tilde{A} \in \mathrm{Hom}_G(Z, \mathrm{Ind}_K^G(W))$. It is easy to check that the corresondence $A \to \tilde{A}$ gives rise to an isomorphism in (3.1). (The inverse is just precomposition by evaluation at the identity element $1 \in G$.) $\checkmark$

REMARK: All the previous considerations apply to complex modules.

PROBLEM: Let V be a complex class 1 module with respect to the symmetric pair (G, K) of compact type. Show that the fixed point set of K in V is 1-dimesional. (Hint: Consider $\mathrm{Ind}_K^G(1)$, where 1 is the 1-dimensional trivial module and use (3.2) along with the remark before Theorem 2.2.)

71

COROLLARY 3.2. *Let V be a finite dimensional orthogonal G-module . Given a K-submodule W_o of S^2V, define*

(3.3)
$$W = \mathrm{span}\,(G \cdot W_o) \subset S^2V.$$

Let $\bar{W}$ be the sum of those irreducible G-submodules of S^2V that when restricted to K contain an irreducible component of W_o. Then we have

(3.4)
$$W \subset \bar{W}.$$

PROOF: Let $p : S^2V \to W_o$ denote the orthogonal projection; $p \in \mathrm{Hom}_K(S^2V, W_o)$. For $C \in S^2V$, define the map $\Psi(C) : G \to W_o$ by $\Psi(C)(a) = p(a \cdot C)$, $a \in G$. Then $\Psi(C) \in \mathrm{Ind}_K^G(W_o)$ so that we obtain a map $\Psi : S^2V \to \mathrm{Ind}_K^G(W_o)$ that is actually a homomorphism of G-modules. Clearly, $\ker \Psi = W^\perp$ so that $\mathrm{im}\,\Psi \cong W \subset \mathrm{Ind}_K^G(W_o)$ as G-modules. Using (3.1), we have

$$\dim \mathrm{Hom}_G(\bar{W}^\perp, W) \leq \dim \mathrm{Hom}_G(\bar{W}^\perp, \mathrm{Ind}_K^G(W_o))$$
$$= \dim \mathrm{Hom}_K(\bar{W}^\perp|_K, W_o) = 0$$

and (3.4) follows. $\checkmark$

For $W_o \subset S^2V$ a trivial K-submodule, a converse of Corollary 3.2 is given by the following:

THEOREM 3.3. *Let $M = G/K$ be a compact isotropy irreducible Riemannian homogeneous space and V a class 1 module with respect to the pair (G, K). Assume that $V|_K$ has multiplicity 1 decomposition into irreducible components. Setting $W_o = \mathbf{R} \cdot \mathrm{proj}\,[v^0] \subset S^2V$, where v^0 is a K-fixed unit vector in V, the G-module W given in (3.3) contains all class 1 submodules of S^2V with respect to (G, K).*

PROOF: Let $f = f_V : M \to S_V$ be given by (2.1). By Theorem 2.2, f is a full minimal immersion with induced naturally reductive Riemannian metric on M. f

72

is equivariant with respect to the homomorphism $\rho : G \to O(V)$ that defines the G-module structure on V.

In what follows, taking higher order derivatives of f, we will define higher fundamental forms of f (cf. [Wallach]). For l a positive integer, we consider the Riemannian connected vector bundle

$$(3.5) \qquad S^l(T^*(M)) \otimes f^*T(S_V) = \mathrm{Hom}\,(S^l(T(M)), f^*T(S_V)).$$

We note that (3.5) is a G-vector bundle, where the action of G on (3.5) is induced by the actions of G on G/K by left translations and on S_V by ρ. (In general, a vector bundle ξ over $M = G/K$ is said to be a G-vector bundle if G acts on (the total space of) ξ by vector bundle isomorphisms and each $a \in G$ induces on the base G/K the left translation by a. In this case, a C^∞-section σ of ξ is said to be G-invariant if $\sigma_{a \cdot x} = a \cdot \sigma_x$ holds for $x \in M$ and $a \in G$. (Note that a special case of these concepts have already been used in Problem* after Example 2.1.))

The l-th fundamental form $\beta_l(f)$ of f will be a G-invariant section of (3.5). As a homomorphism, its image O^l is then automatically a G-vector subbundle of $f^*T(S_V)$. The way $\beta_l(f)$ will be defined below will show that, for some positive integer k, we will have the (fibrewise) orthogonal decomposition

$$(3.6) \qquad\qquad f^*T(S_V) = O^1 \oplus \ldots O^k$$

into G-vector subbundles. k is said to be the $degree$ of f; for $1 \leq l \leq k$, the G-vector subbundle

$$O^1 \oplus \ldots \oplus O^l \subset f^*T(S_V)$$

is said to be the l-th osculating vector bundle of f.

We now define the higher fundamental forms of f by induction with respect to l. For $l = 1$, we set $\beta_1(f) = f_*$ and hence $O^1 = T(M)$, where the latter is considered to be a G-vector subbundle of $f^*T(S_V)$. In fact, we have the orthogonal decomposition

$$f^*T(S_V) \cong T(M) \oplus \nu_f,$$

73

where ν_f is the normal bundle of f.

For $l = 2$, we set $\beta_2(f) = \beta(f)$ the second fundamental form of f introduced in §1 of Chapter I. We define O^2 as the image of $\beta_2(f)$. Clearly, O^2 is a G-vector subbundle of ν_f.

In general, we define inductively

$$\beta_l(f)(X^1, \ldots, X^l) = (\nabla_{X^1} \beta_{l-1}(f))(X^2, \ldots, X^l)^{\perp},$$

where $X^1, \ldots, X^l$ are vector fields on M and $\perp$ denotes the orthogonal projection onto $(O^1 \oplus \ldots \oplus O^{l-1})^{\perp} \subset f^*T(S_V)$. Due to the fact that S_V has a constant curvature, $\beta_l(f)$ can be shown to be symmetric in all its arguments. As above, we define O^l as the image of $\beta_l(f)$.

Finally, let $\beta_k(f)$ be the last nonzero higher fundamental form of f. Now, (3.6) follows by an easy argument using that all data are analytic.

Returning to the proof we restrict (3.6) to the base point $o \in M$ to obtain the orthogonal decomposition

(3.7) $$V|_K \cong O_o^0 \oplus O_o^1 \oplus \ldots \oplus O_o^k,$$

into K-submodules, where $O_o^0 = \mathbf{R} \cdot v^0$ is the trivial K-module as v^0 is left fixed by K. (3.7) holds because the fibre of $f^*T(S_V)$ over o, as a K-module, is nothing but $T_o(S_V) \cong (v^0)^{\perp} \subset V$. By hypothesis, $V|_K$ has multiplicity 1 decomposition into irreducible components so that, for $l \neq l'$, $l, l' = 0, \ldots, k$, we have

(3.8) $$\mathrm{Hom}_K(O_o^l, O_o^{l'}) = \{0\}.$$

We now claim that if $A \in \mathrm{Hom}_K(V, V)$ is a symmetric positive definite endomorphism of V satisfying

(3.9) $$A(G \cdot v^0) \subset S_V$$

74

then A is the identity. First, (3.8) implies that A leaves O_o^l invariant, $l = 0, \ldots, k$. We show by induction with respect to l that $A|_{O_o^l}$ is the identity. For $l = 0$, Av^0 is a K-fixed vector of norm 1 (by (3.9)) so that $Av^0 = v^0$ since A is positive definite. Thus A is the identity on O_o^0. Assume now that A is the identity on $O_o^0 \oplus \ldots \oplus O_o^{l-1}$ for some $(1 \le)l \le k$.

Let $c : \mathbf{R} \to M$ be a nontrivial geodesic with $c(0) = o$ and define

$$\gamma^{(j)}(0) = \frac{d^j f(c(t))}{dt^j}\Big|_{t=0} \in V, \ j \ge 0.$$

By the definition of osculating spaces, we have

(3.10) $$\gamma^{(j)}(0) \in O_o^0 \oplus O_0^1 \oplus \ldots \oplus O_o^j$$

with nonzero component in O_o^j. In fact, these components, for various c's, span O_0^j so that to perform the induction step it is enough to show that

$$|A\gamma^{(l)}(0)|^2 = |\gamma^{(l)}(0)|^2.$$

By (3.9), we have

$$0 = \frac{d^{2l}(|A\gamma(t)|^2)}{dt^{2l}}\Big|_{t=0} = \sum_{j=0}^{2l} \binom{2l}{j} \langle A\gamma^{(j)}(0), A\gamma^{(2l-j)}(0) \rangle$$

so that

(3.11) $$|A\gamma^{(l)}(0)|^2 = \sum_{j=0}^{l-1} a_j^l \langle A\gamma^{(j)}(0), A\gamma^{(2l-j)}(0) \rangle,$$

where $a_j^l = -2\binom{2l}{j}/\binom{2l}{l}$ is independent of A. In particular, (3.11) holds for $A = I_V$. For fixed $0 \le j \le l - 1$, we use (3.10) to decompose

$$\gamma^{(2l-j)} = \sum_{i=0}^{2l-j} w_i, \ w_i \in O_0^i.$$

75

Using the induction hypothesis (twice), we have

$$\langle A\gamma^{(j)}(0), A\gamma^{(2l-j)}(0)\rangle = \langle \gamma^{(j)}(0), A\gamma^{(2l-j)}(0)\rangle$$

$$= \sum_{i=0}^{2l-j}\langle \gamma^{(j)}(0), Aw_i\rangle = \sum_{i=0}^{l-1}\langle \gamma^{(j)}(0), Aw_i\rangle$$

$$= \sum_{i=0}^{l-1}\langle \gamma^{(j)}(0), w_i\rangle = \sum_{i=0}^{2l-j}\langle \gamma^{(j)}(0), \gamma^{(2l-j)}(0)\rangle.$$

Substituting this into (3.11) we obtain

$$|A\gamma^{(l)}(0)|^2 = \sum_{j=0}^{l-1} a_j^l \langle \gamma^{(j)}(0), \gamma^{(2l-j)}(0)\rangle = |\gamma^{(l)}(0)|^2$$

and the induction step is complete. The claim follows.

To finish the proof of the theorem, let Z be an irreducible G-submodule of $S^2 V$ that is class 1 with respect to (G, K) and assume that Z is *not* contained in W. We may then assume that Z is orthogonal to W since otherwise we apply to Z the orthogonal projection of $S^2 V$ onto $W^\perp$ that is a homomorphism of G-modules. Let $C \in Z$ be left fixed by K. Choose $\varepsilon > 0$ such that $I + \varepsilon C > 0$. By orthogonality, we have

$$\langle I + \varepsilon C, a \cdot \operatorname{proj}[v^0]\rangle = \langle I, a \cdot \operatorname{proj}[v^0]\rangle = \langle I, \operatorname{proj}[v^0]\rangle = |v^0|^2 = 1$$

so that, setting $A = \sqrt{I + \varepsilon C}$ we obtain a symmetric positive definite endomorphism $A \in \operatorname{Hom}_K(V, V)$ which satisfies (3.9). By the above, $A = I_V$. Playing this back to C, we conclude that $C = 0$ and the proof is complete. $\checkmark$

PROBLEM *: Let $M = G/K$ be a compact Riemannian homogeneous space and $f : M \to S_V$ a full isometric minimal immersion. Show that if f is of degree ≤ 3 then $\mathcal{M}_f = \{0\}$ (cf. [DoCarmo-Wallach][Wallach]). (Hint: Let $c : \mathbf{R} \to M$ be a nontrivial geodesic and

$$\gamma^{(j)}(0) = \frac{d^j f(c(t))}{dt^j}\Big|_{t=0} \in V, \ j = 0, 1, 2, 3.$$

Show, by differentiating, that, for $j = 0, 1, 2, 3$, we have $\gamma(0) \cdot \gamma^{(j)}(0) \in \mathcal{Z}_f (\supset \mathcal{W}_f)$. Finally, make the scalar product of these with an arbitrary symmetric endomorphism $C \in \mathcal{F}_f$ to conclude that $C = 0$.)

We now return to standard eigenmaps. Let $M = G/K$ be a Riemannian homogeneous space, λ an eigenvalue of the Laplace-Beltrami operator Δ^M with eigenspace $V_\lambda \subset C^\infty(M)$ and $f_\lambda : M \to S^{n(\lambda)}$ a standard λ-eigenmap whose components comprise an orthonormal basis $\{f_\lambda^i\}_{i=0}^{n(\lambda)} \subset V_\lambda$ with respect to the normalized L^2-scalar product in (2.4) on V_λ. We denote by $\rho_\lambda : G \to O(V_\lambda)$ the homomorphism associated with the G-module structure on $V_\lambda \cong \mathbf{R}^{n(\lambda)+1}$ with respect to which f_λ is equivariant. Setting $\mathcal{W}_\lambda = \mathcal{W}_{f_\lambda} \subset S^2(V_\lambda)$ and $\mathcal{E}_\lambda = \mathcal{W}_\lambda^\perp$, by equivariance of f_λ, (2.3) of Chapter I translates into

$$(3.12) \qquad \mathcal{W}_\lambda = \operatorname{span} \{ \operatorname{proj} [\rho_\lambda(a) \cdot f_\lambda(o)] | a \in G \} = \operatorname{span} (G \cdot \operatorname{proj} [f_\lambda(o)]).$$

We are now in the position to apply Corollary 3.2. The cast is $V = V_\lambda$ and $W_o = \mathbf{R} \cdot \operatorname{proj} [f_\lambda(o)]$. Clearly, W_o is a trivial K-submodule of $S^2(V_\lambda)$. Comparing (3.3) and (3.12) we readily see that W in (3.3) specializes to $\mathcal{W}_\lambda$. We obtain the following:

PROPOSITION 3.4. *Let $\bar{\mathcal{E}}_\lambda$ denote the sum of those G-submodules of $S^2(V_\lambda)$ that are not class 1 with respect to (G, K). Then we have*

$$(3.13) \qquad\qquad \bar{\mathcal{E}}_\lambda \subset \mathcal{E}_\lambda. \checkmark$$

REMARK: We could have obtained (3.13) more directly by taking an irreducible component Z of $\mathcal{W}_\lambda$ and arguing that, for some $a \in G$, $\operatorname{proj} [f_\lambda(a \cdot o)]$ is not orthogonal to Z and hence projects down to Z yielding a nonzero vector in Z which is left fixed by $aKa^{-1} \subset G$. We have, however, chosen a somewhat more abstract approach since, due to its universal features, it will be applied to various other instances subsequently.

Assume now that $M = G/K$ is isotropy irreducible so that $f_\lambda : M \to S^{n(\lambda)}$ is a standard minimal immersion. Setting $\mathcal{Z}_\lambda = \mathcal{Z}_{f_\lambda} \subset S^2(V_\lambda)$ and $\mathcal{F}_\lambda = \mathcal{Z}_\lambda^\perp$, by equivariance of f_λ, (2.13) of Chapter I translates into

$$\mathcal{Z}_\lambda = \operatorname{span} \{ \operatorname{proj} [\rho_\lambda(a) \cdot (f_\lambda)_* X_o^\cdot] | X_o \in T_o(M) \}$$

(3.14) $$= \operatorname{span} (G \cdot \{ \operatorname{proj} [(f_\lambda)_* X_o^\cdot] | X_o \in T_o(M) \}).$$

This time the cast for Corollary 3.2 is $V = V_\lambda$ and $W_o = \operatorname{span} \{ \operatorname{proj} [(f_\lambda)_* X_o^\cdot] | X_o \in T_o(M) \}$. We observe that, by equivariance of f_λ (with respect to $\rho_\lambda|_K$) and by (2.2) of Chapter I, $W_o \cong S^2(T_o(M))$ as K-modules. A quick comparison of (3.3) and (3.14) convinces us that W in (3.3) specializes now to $\mathcal{F}_\lambda$. Hence we have the following:

PROPOSITION 3.5. Let $\bar{\mathcal{F}}_\lambda$ denote the sum of those irreducible G-submodules of $S^2(V_\lambda)$ that when restricted to K do not contain any irreducible component of $S^2(T_o(M))$. Then we have

(3.15) $$\bar{\mathcal{F}}_\lambda \subset \mathcal{F}_\lambda. \checkmark$$

The importance of Propositions 3.4 and 3.5 lies in the fact that the dimensions of $\bar{\mathcal{E}}_\lambda$ and $\bar{\mathcal{F}}_\lambda$ can be determined using representation theory (since they are given by representation theoretical data) so that, using (3.13) and (3.15), we obtain lower estimates for the dimensions of the moduli spaces $\mathcal{L}_\lambda$ and $\mathcal{M}_\lambda$. We will pursue this program for the rest of this chapter in two special cases of rank 1 symmetric spaces. Let $M = S^m \subset \mathbf{R}^{m+1}$ be the Euclidean m-sphere with base point $o = (1, 0, \ldots, 0) \in S^m$ and $G = SO(m+1)$. The isotropy subgroup $K = SO(m)(= [1] \times SO(m) \subset SO(m+1))$ acts on $T_o(S^m) \cong \mathbf{R}^m$ by ordinary matrix multiplication so that the action is transitive on the unit sphere. We obtain that $S^m = SO(m+1)/SO(m)$ is an isotropy irreducible homogeneous space. The Riemannian metric of S^m induced from $\mathbf{R}^{m+1}$ being Riemannian homogeneous is naturally reductive. It is well-known that the k-th eigenvalue of the Laplace-Beltrami operator Δ^{S^m} is given by

(3.16) $$\lambda_k = k(k + m - 1),\ k \geq 0,$$

and the eigenspace V_{λ_k} associated to λ_k is nothing but the linear space $\mathcal{H}^k_{S^m}(\mathbf{R})$ of real spherical harmonics of order k on S^m. Equivalently, $\mathcal{H}^k_{S^m}(\mathbf{R})$ consists of the restrictions (to S^m) of all harmonic homogeneous polynomials of degree k on $\mathbf{R}^{m+1}$. (To verify this we only have to compare the Laplace-Beltrami operators Δ^{S^m} and $\Delta^{\mathbf{R}^{m+1}}$.) As an $SO(m + 1)$-module, $\mathcal{H}^k_{S^m}(\mathbf{R})$ is irreducible. Computation shows that

(3.17) $$\dim \mathcal{H}^k_{S^m}(\mathbf{R}) = n(\lambda_k) + 1 = (m + 2k - 1)\frac{(m + k - 2)!}{k!(m - 1)!}.$$

The choice of an orthonormal basis $\{f^i_{\lambda_k}\}^{n(\lambda_k)}_{i=0} \subset \mathcal{H}^k_{S^m}(\mathbf{R})$ with respect to the normalized L^2-scalar product (2.4) defines, as usual, a standard minimal immersion

$$f_{\lambda_k} : S^m \to S^{n(\lambda_k)}.$$

An orthonormal basis is given by the Gegenbauer polynomials (cf. [Vilekin]). Note that the case $k = 1$ is uninteresting since $f_{\lambda_1} : S^m \to S^m$ is nothing but an isometry.

The case $k = 2$ classical; the standard minimal immersion

$$f_{\lambda_2} : S^m \to S^{\frac{m(m+3)}{2}-1}$$

is called the Veronese map. In the Remark after Theorem 4.1 of Chapter I we mentioned $f_{\lambda_2} : S^2 \to S^4$ whose image is a minimally imbedded real projective space $\mathbf{R}P^2$ known as the classical Veronese surface in S^4. (Actually, it can also be thought of as a singular variety of a homogeneous isoparametric family of hypersurfaces of degree 3 on S^4; the action is given by $\rho_{\lambda_2} : SO(4) \to SO(\mathbf{R}^5)$. Note also that, by Problem** at the end of §2, the unique minimal homogeneous hypersurface is $SO(4)/\mathbf{Z}_2 \times \mathbf{Z}_2$ on which the induced Riemannian homogeneous metric is not naturally reductive.) In §4 of Chapter I we dealt with the standard minimal immersion $f_{\lambda_2} : S^3 \to S^8$ whose explicit form is given there in (4.6).

PROBLEM: Show that a standard minimal immersion $f_{\lambda_2} : S^m \to S_{\mathcal{H}_{S^m}^k(\mathbf{R})}$ can be written as

$$f_{\lambda_2}(x_0, \ldots, x_m) = \frac{1}{I-J} \sum_{k=0}^{m} (x_k^2 - \frac{1}{m+1})\varphi_k$$
$$+ \frac{2}{J} \sum_{0 \le i < j \le m} x_i x_j \varphi_{ij}, \ (x_0, \ldots, x_m) \in S^m \subset \mathbf{R}^{m+1},$$

where the quadratic polynomials φ_k and φ_{ij} are given by

$$\varphi_k(x_0, \ldots, x_m) = x_k^2$$

and

$$\varphi_{ij}(x_0, \ldots, x_m) = x_i x_j$$

and $I = |\varphi_k|^2$ and $J = |\varphi_{ij}|^2$.

We wish to determine $\mathcal{E}_{\lambda_k} \subset S^2(\mathcal{H}_{S^m}^k(\mathbf{R}))$ as an $SO(m+1)$-module. By Proposition 3.4, we have

(3.18) $$\bar{\mathcal{E}}_{\lambda_k} \subset \mathcal{E}_{\lambda_k},$$

where $\bar{\mathcal{E}}_{\lambda_k}$ denotes the sum of those $SO(m+1)$-submodules of $S^2(\mathcal{H}_{S^m}^k(\mathbf{R}))$ that are class 1 with respect to $(SO(m+1), SO(m))$. Moreover, as a special case of the Branching Rule discussed in the next sextion, we have

(3.19) $$\mathcal{H}_{S^m}^k(\mathbf{R})|_{SO(m)} \cong \sum_{l=0}^{k} \mathcal{H}_{S^{m-1}}^l(\mathbf{R})$$

so that all the hypotheses of Theorem 3.3 are satisfied (for $(SO(m+1), SO(m))$ and $V = \mathcal{H}_{S^m}^k(\mathbf{R})$). We obtain that equality holds in (3.18).

80

THEOREM 3.6. *For the standard minimal immersion*

$$f_{\lambda_k} : S^m \to S^{n(\lambda_k)}, \ S^m = SO(m+1)/SO(m),$$

$\mathcal{E}_{\lambda_k}$ *is the sum of those $SO(m+1)$-submodules of $S^2(\mathcal{H}^k_{S^m}(\mathbf{R}))$ that are not class 1 with respect to $(SO(m+1), SO(m))$.* $\checkmark$

PROBLEM*: Show that

$$O^l_o \cong \mathcal{H}^l_{S^{m-1}}(\mathbf{R}), \ l = 0, 1, \ldots, k,$$

in particular, the degree of f_{λ_k} is k (cf. [DoCarmo-Wallach]). (Hint: $\beta_l(f_{\lambda_k})$: $S^l(\mathbf{R}^m) \to O^l_o$ is an $SO(m)$-module epimorphism, where $\mathbf{R}^m$ is the standard $SO(m)$-module given by matrix multiplication.)

We now turn to minimal immersions.

THEOREM 3.7. *We have*

$$\bar{\mathcal{F}}_{\lambda_k} \subset \mathcal{F}_{\lambda_k},$$

where $\bar{\mathcal{F}}_{\lambda_k}$ is the sum of those irreducible $SO(m+1)$-submodules of $S^2(\mathcal{H}^k_{S^m}(\mathbf{R}))$ that when restricted to $SO(m)$ do not contain the trivial $SO(m)$-module $\mathcal{H}^0_{S^{m-1}}(\mathbf{R})$ and $\mathcal{H}^2_{S^{m-1}}(\mathbf{R})$.

PROOF: This follows from Proposition 3.5 since as $SO(m)$-modules

$$S^2(\mathbf{R}^m) \cong S^2(\mathcal{H}^1_{S^{m-1}}(\mathbf{R})) \cong \mathcal{H}^0_{S^{m-1}}(\mathbf{R}) \oplus \mathcal{H}^2_{S^{m-1}}(\mathbf{R}). \checkmark$$

Now let $M = \mathbf{C}P^m$ with base point $o = [1 : 0 : \ldots : 0] \in \mathbf{C}P^m$ and $G = U(m+1)$. The isotropy subgroup K is generated by the center

$$S^1 = \{\, \mathrm{diag}\,(e^{i\theta}, \ldots, e^{i\theta}) \in U(m+1) | \theta \in \mathbf{R}\}$$

and $U(m)(= [1] \times U(m)) \subset U(m+1)$ so that it coincides with $U(1) \times U(m)$. Thus, we have $\mathbf{C}P^m = U(m+1)/U(1) \times U(m)$. Moreover, $U(m)$ acts on $T_o(\mathbf{C}P^m) \cong \mathbf{C}^m$ by

81

matrix multiplication and is, in particular, transitive on the unit sphere of $T_o(\mathbf{C}P^m)$. We obtain that $\mathbf{C}P^m$ is an isotropy irreducible Riemannian homogeneous space. The unitary group $U(m+1)$ acts on the unit sphere $S^{2m+1} \subset \mathbf{C}^{m+1}$ transitively and the isotropy subgroup at $\bar{o} = (1, 0, \ldots, 0) \in \mathbf{C}^{m+1}$ is nothing but $U(m)$. Note that S^{2m+1} as $U(m+1)/U(m)$ is not isotropy irreducible.

The Hopf map $H : S^{2m+1} \to \mathbf{C}P^m$ corresponds to the projection $U(m+1)/U(m) \to \mathbf{C}P^m$ induced by the action of S^1. With respect to the standard Riemannian metric on S^{2m+1} which is naturally reductive (with respect to both $SO(2m+2)$ and $U(m+1)$) the induced Riemannian metric on $\mathbf{C}P^m$ is also naturally reductive so that (1.5) applies to the Riemannian submersion H. Note that the Riemannian metric induced on $\mathbf{C}P^m$ is such that the sectional curvatures vary between $1/4$ and 1.

Let μ be an eigenfunction of $\triangle^{\mathbf{C}P^m}$ with eigenvalue λ. Then, by (1.5), $\bar{\mu} = \mu \circ \pi$ is also an eigenfunction of $\triangle^{S^{2m+1}}$ with the same eigenvalue λ. By the above, $\lambda = \lambda_k$, where λ_k is given in (3.16), and $\bar{\mu}$ is a harmonic homogeneous polynomial of degree k on $\mathbf{R}^{2m+2} = \mathbf{C}^{m+1}$.

In general, using complex coordinates $(z_0, \ldots, z_m) \in \mathbf{C}^{m+1}$, a complex valued spherical harmonic of degree k can uniquely be written as the sum of harmonic homogeneous polynomials of bidegree (p, q), $p + q = k$, i.e. of degree p in $z_0, \ldots, z_m$ and of degree q in $\bar{z}_0, \ldots, \bar{z}_m$. We denote by $\mathcal{H}^{p,q}_{\mathbf{C}P^m}$ the complex vector space of harmonic homogeneous polynomials of bidegree (p, q) on $\mathbf{C}P^m$. In fact $\mathcal{H}^{p,q}_{\mathbf{C}P^m}(\subset C^\infty(\mathbf{C}P^m) \otimes_\mathbf{R} \mathbf{C})$ is an irreducible $U(m+1)$-module and we have the decomposition

$$(3.20) \qquad \mathcal{H}^k_{S^{2m+1}}|_{U(m+1)} \cong \sum_{\substack{p+q=k \\ p,q \geq 0}} \mathcal{H}^{p,q}_{\mathbf{C}P^m},$$

where $\mathcal{H}^k_{S^{2m+1}}$ is the complexification of $\mathcal{H}^k_{S^{2m+1}}(\mathbf{R})$. (Note that we have already discussed these on account of the complex Veronese surface in §4 of Chapter I.) By definition, the center $S^1 \subset U(m+1)$ acts on $\mathcal{H}^{p,q}_{\mathbf{C}P^m}$ by weight $p - q$.

Returning to our special case, we see that $\bar{\mu}$ is S^1-invariant so that $k = 2p$ and $\bar{\mu} \in \mathcal{H}^{p,p}_{\mathbf{C}Pm}$. Moreover, $\bar{\mu}$ is *real valued* so that it is contained in the *real* irreducible $U(m+1)$-module $\mathcal{H}^{p,p}_{\mathbf{C}Pm}(\mathbf{R})$ of real valued harmonic homogeneous polynomials of bidegree (p,p) on $\mathbf{C}^{m+1}$; $\mathcal{H}^{p,p}_{\mathbf{C}Pm}(\mathbf{R})$ is actually a real form of $\mathcal{H}^{p,p}_{\mathbf{C}Pm}$.

Summarizing we obtain that the p-th eigenvalue of the Laplace-Beltrami operator $\triangle^{\mathbf{C}P^m}$ is

$$\lambda_p = 4p(p+m), \; p \geq 0,$$

and the corresponding (real) eigenspace V_{λ_p} is $\mathcal{H}^{p,p}_{\mathbf{C}Pm}(\mathbf{R})$. Inspection shows that

$$\dim \mathcal{H}^{p,p}_{\mathbf{C}Pm}(\mathbf{R}) = n(\lambda_p) + 1 = \binom{m+p}{p}^2 - \binom{m+p-1}{p-1}^2.$$

The choice of an orthonormal basis $\{f^i_{\lambda_p}\}^{n(\lambda_p)}_{i=0} \subset \mathcal{H}^{p,p}_{\mathbf{C}Pm}(\mathbf{R})$ with respect to the normalized L^2-scalar product (2.4) defines a standard minimal immersion

(3.21) $$f_{\lambda_p} : \mathbf{C}P^m \to S^{n(\lambda_p)}.$$

Again we are interested in the $U(m+1)$-module $\mathcal{E}_{\lambda_p} \subset S^2(\mathcal{H}^{p,p}_{\mathbf{C}Pm}(\mathbf{R}))$. Restricting to $U(m)$, we obtain

(3.22) $$\mathcal{H}^{p,p}_{\mathbf{C}Pm}|_{U(m)} \cong \sum_{0 \leq r,s \leq p} \mathcal{H}^{r,s}_{\mathbf{C}Pm-1}$$

and so Proposition 3.4 and Theorem 3.3 apply. ($\mathcal{H}^{p,p}_{\mathbf{C}Pm}(\mathbf{R})|_{U(m)}$ has multiplicity 1 decomposition into irreducible components since the same holds for the complexification. Note also that $U(1)$ acts as a character on each of the components so that it has no effect on irreducibility.)

THEOREM 3.8. *For the standard minimal immersion*

$$f_{\lambda_p} : \mathbf{C}P^m \to S^{n(\lambda_p)}, \; \mathbf{C}P^m = U(m+1)/U(1) \times U(m),$$

as a $U(m+1)$-module, $\mathcal{E}_{\lambda_p}$ is the sum of those $U(m+1)$-submodules of $S^2(\mathcal{H}_{\mathbf{C}P^m}^{p,p}(\mathbf{R}))$ that are not class 1 with respect to $(U(m+1), U(1) \times U(m))$. $\checkmark$

As in the spherical case, we have the lower estimate

$$\bar{\mathcal{F}}_{\lambda_p} \subset \mathcal{F}_{\lambda_p},$$

where $\bar{\mathcal{F}}_{\lambda_p}$ is the sum of those irreducible $U(m+1)$-submodules of $S^2(\mathcal{H}_{\mathbf{C}P^m}^{p,p}(\mathbf{R}))$ that when restricted to $U(m)$ do not contain any irreducible component of $S^2(\mathbf{C}^m)$ viewed as a *real* $U(m)$-module. This can be conveniently rephrased by saying that the complexification $\bar{\mathcal{F}}_{\lambda_p} \otimes_{\mathbf{R}} \mathbf{C}$ is the sum of those complex irreducible $U(m+1)$-submodules of $S^2(\mathcal{H}_{\mathbf{C}P^m}^{p,p}(\mathbf{R})) \otimes_{\mathbf{R}} \mathbf{C}$ that when restricted to $U(m)$ do not contain any irreducible component isomorphic to

$$\mathcal{H}_{\mathbf{C}P^{m-1}}^{0,0}, \mathcal{H}_{\mathbf{C}P^{m-1}}^{2,0}, \mathcal{H}_{\mathbf{C}P^{m-1}}^{1,1}, \text{ and } \mathcal{H}_{\mathbf{C}P^{m-1}}^{0,2}.$$

This is because, we have

$$S^2(\mathbf{C}^m) = S^2(\mathbf{R}^{2m}) = S^2(\mathcal{H}_{S^{2m-1}}^1(\mathbf{R})) \cong \mathcal{H}_{S^{2m-1}}^0(\mathbf{R}) \oplus \mathcal{H}_{S^{2m-1}}^2(\mathbf{R})$$

so that complexifying and restricting to $U(m) \subset SO(2m)$, by (3.22), we have

$$\mathcal{H}_{S^{2m-1}}^0|_{U(m)} \cong \mathcal{H}_{\mathbf{C}P^{m-1}}^{0,0}$$

and

$$\mathcal{H}_{S^{2m-1}}^2|_{U(m)} \cong \mathcal{H}_{\mathbf{C}P^{m-1}}^{2,0} \oplus \mathcal{H}_{\mathbf{C}P^{m-1}}^{1,1} \oplus \mathcal{H}_{\mathbf{C}P^{m-1}}^{0,2}.$$

THEOREM 3.9. *For the standard minimal immersion*

$$f_{\lambda_p} : \mathbf{C}P^m \to S^{n(\lambda_p)}, \quad \mathbf{C}P^m = U(m+1)/U(1) \times U(m),$$

the $U(m+1)$-module $\mathcal{F}_{\lambda_p}$ contains a $U(m+1)$-submodule $\bar{\mathcal{F}}_{\lambda_p}$ whose complexification $\bar{\mathcal{F}}_{\lambda_p} \otimes_{\mathbf{R}} \mathbf{C}$ is the sum of those complex irreducible $U(m+1)$-submodules of

84

$S^2(\mathcal{H}^{p,p}_{\mathbf{C}Pm}) = S^2(\mathcal{H}^{p,p}_{\mathbf{C}Pm}(\mathbf{R})) \otimes_{\mathbf{R}} \mathbf{C}$ that, when restricted to $U(m)$, do not contain any irreducible component isomorphic to

$$\mathcal{H}^{0,0}_{\mathbf{C}Pm-1}, \ \mathcal{H}^{2,0}_{\mathbf{C}Pm-1}, \ \mathcal{H}^{1,1}_{\mathbf{C}Pm-1} \ \text{and} \ \mathcal{H}^{0,2}_{\mathbf{C}Pm-1}.\checkmark$$

In §4 of Chapter III we will derive a complete decomposition of $S^2(\mathcal{H}^{p,p}_{\mathbf{C}Pm})$ into irreducible $U(m+1)$-submodules. As in the spherical case, by Theorem 3.8, this will immediately yield the complete decomposition of $\mathcal{E}_{\lambda_p} \otimes_{\mathbf{R}} \mathbf{C}$. (Note that some components of $\mathcal{E}_{\lambda_p} \otimes_{\mathbf{R}} \mathbf{C}$ have previously been found by [Urakawa;2].) In particular, $\dim \mathcal{L}_{\lambda_p}(= \dim \mathcal{E}_{\lambda_p})$ will follow using the Weyl dimension formula. For minimal immersions, Theorem 3.9 along with branching will give the complete decomposition of $\bar{\mathcal{F}}_{\lambda_p} \otimes_{\mathbf{R}} \mathbf{C}$ and thereby a lower bound for $\dim \mathcal{M}_{\lambda_p}$.

PROBLEM*: Show that the degree of the standard minimal immersion in (3.21) is $2p$ and

$$O^l_o \otimes_{\mathbf{R}} \mathbf{C} \cong \sum_{\substack{r+s=l \\ 0 \leq r,s \leq p}} \mathcal{H}^{r,s}_{\mathbf{C}Pm-1}, \ l = 0, 1, \ldots, 2p.$$

as $U(m)$-modules (cf. [Mashimo;1,2]).

REMARK: While the main objective of these notes is to study the cases $M = S^m$ and $M = \mathbf{C}P^m$, we see no harm in noting that for remaining compact rank 1 symmetric spaces, namely, for the quaternionic projective space $M = \mathbf{H}P^m = Sp(m+1)/Sp(1) \cdot Sp(m)$ and the Cayley projective plane $M = CaP^2 = F_4/Spin(9)$ the degree of the standard minimal immersion f_{λ_p} is $2p$. For further information about the osculating bundles of f_{λ_p} and the geometry of the helical geodesic immersions in general consult with [Sakamoto][Tsukada][Mashimo;1,2].

§4. Decomposition of $\mathcal{E}_{\lambda_k}$ and $\mathcal{F}_{\lambda_k}$ associated with a standard minimal immersion $f_{\lambda_k} : S^m \to S^{n(\lambda_k)}$

Theorem 3.6 says that the $SO(m+1)$-module $\mathcal{E}_{\lambda_k} \subset S^2(\mathcal{H}^k_{S^m}(\mathbf{R}))$ associated to a standard minimal immersion $f_{\lambda_k} : S^m \to S^{n(\lambda_k)}$ is obtained from $S^2(\mathcal{H}^k_{S^m}(\mathbf{R}))$ by discarding those irreducible submodules that when restricted to $SO(m)$ contain the trivial $SO(m)$-module. Thus, to determine (the irreducible components of) $\mathcal{E}_{\lambda_k}$ we have to know how does an irreducible $SO(m+1)$-module split when restricted to $SO(m)$. This happens to be well-known and is the content of the Branching Rule stated below. In addition, and this is a harder task, we have to derive the complete decomposition of $S^2(\mathcal{H}^k_{S^m}(\mathbf{R}))$ into irreducible components. This will be stated in this section and the fairly technical proof, due to [Wallach], will be given in the next (final) section of this chapter.

To begin with we recall a few facts from representation theory of the special orthogonal group (the proofs can be found in any standard textbook on representation theory, cf. [Börner][Humphreys][Naimark-Stern][Zhelobenko]). For convenience we work with complex modules; this is certainly no loss of generality since the real irreducible modules we encounter here with will remain, after complexification, irreducible (as complex modules).

Let $T \subset SO(m+1)$ be the standard maximal torus. Its dimension is $\ell = ||\frac{m+1}{2}||$, where $||[x]||$ denotes the greatest integer $\leq x$. In fact, if

$$R_\theta = \begin{bmatrix} \cos\theta & \sin\theta \\ & \\ -\sin\theta & \cos\theta \end{bmatrix} \in SO(2)$$

86

denotes the (clockwise) rotation of the plane by angle $\theta \in \mathbf{R}$, then we have

$$T = \{R_{\theta_1} \times \ldots \times R_{\theta_\ell} \times [1] | \theta_j \in \mathbf{R},\, j = 1, \ldots, \ell\} \text{ if } m + 1 = 2\ell + 1$$

and

$$T = \{R_{\theta_1} \times \ldots \times R_{\theta_\ell} | \theta_j \in \mathbf{R},\, j = 1, \ldots, \ell\} \text{ if } m + 1 = 2\ell.$$

These representations establish an isomorphism of T with $\mathbf{R}^\ell/(2\pi\mathbf{Z})^\ell$ that we think of as identification in the sequel.

Now let V be a finite dimensional complex $SO(m+1)$-module. By restriction, V is also a T-module. In other words, T is a compact commutative Lie group of complex linear automorphisms of V. A simultaneous eigenvector $v \in V$ of all elements of T is said to be a *weight vector*. If $v \neq 0$ then there exists a unique integral ℓ-tuple $\phi = (\phi_1, \ldots, \phi_\ell) \in \mathbf{Z}^\ell$ such that

$$(\theta_1, \ldots, \theta_\ell) \cdot v = \exp\Big(i \sum_{j=1}^{\ell} \phi_j \theta_j\Big)v, \ (\theta_1, \ldots, \theta_\ell) \in T.$$

Then ϕ is said to be a *weight* (associated to the weight vector $v \in V$). The set of weight vectors with weight ϕ is a linear subspace V_ϕ of V on which $(\theta_1, \ldots, \theta_\ell) \in T$ acts as multiplication by $\exp\big(i \sum_{j=1}^{\ell} \phi_j \theta_j\big)$. The set Φ of weights is finite and we have the direct sum decomposition

$$V = \sum_{\phi \in \Phi} V_\phi.$$

Now consider the (finite) group of all inner automorphisms of $SO(m+1)$ that leave T invariant. Their restrictions to T form what is called the Weyl group $W_{SO(m+1)}$ of $SO(m+1)$. Little inspection reveals that

$$W_{SO(m+1)} \cong S_\ell \triangleright \mathbf{Z}_2^\ell, \text{ if } m + 1 = 2\ell + 1$$

and

$$W_{SO(m+1)} \cong S_\ell \triangleright \mathbf{Z}_2^{\ell-1}, \text{ if } m + 1 = 2\ell,$$

where $\mathcal{S}_\ell$ is the symmetric group on ℓ letters, $\mathbf{Z}_2 = \{\pm 1\}$ and $\triangleright$ denotes semidirect product. For $m+1 = 2\ell+1$, $\sigma \in \mathcal{S}_\ell$ acts on T as it should, namely, as a permutation sending $(\theta_1, \ldots, \theta_\ell)$ into $(\theta_{\sigma(1)}, \ldots, \theta_{\sigma(\ell)})$. Moreover, $(\epsilon_1, \ldots, \epsilon_\ell) \in \mathbf{Z}_2^\ell$ acts by sending $(\theta_1, \ldots, \theta_\ell)$ into $(\epsilon_1 \theta_1, \ldots, \epsilon_\ell \theta_\ell)$. For $m+1 = 2\ell$, the situation is similar, except that by $\mathbf{Z}_2^{\ell-1}$ we mean the subgroup of $\mathbf{Z}_2^\ell$ consisting of those $(\epsilon_1, \ldots, \epsilon_\ell) \in \mathbf{Z}_2^\ell$ for which $\prod_{j=1}^\ell \epsilon_j = 1$. By construction, $W_{SO(m+1)}$ acts on the set of weights Φ. Thus, given a weight ϕ there exists a weight $\rho \in \mathbf{Z}^\ell$ in the Φ-orbit of ϕ satisfying

$$(4.1) \qquad \rho_1 \geq \ldots \geq \rho_{\ell-1} \geq \rho_\ell \geq 0 \text{ if } m+1 = 2\ell+1$$

and

$$(4.2) \qquad \rho_1 \geq \ldots \geq \rho_{\ell-1} \geq |\rho_\ell| \text{ if } m+1 = 2\ell.$$

If V is irreducible, ρ is unique and is actually maximal with respect to the lexicographic order on $\mathbf{Z}^\ell$. Moreover, the weight space V_ρ is 1-dimensional. In this case, we call ρ the *highest weight* of the complex irreducible $SO(m+1)$-module V. It is a fundamental theorem of [É.Cartan] that ρ characterizes V up to equivalence. Moreover, for each ℓ-tuple $\rho \in \mathbf{Z}^\ell$ satisfying (4.1) or (4.2), there is an irreducible complex $SO(m+1)$-module V such that ρ is the highest weight of V. Therefore it is legitimate to denote by $V_{SO(m+1)}^\rho$ the irreducible complex $SO(m+1)$-module with highest weight ρ. In particular, a little effort in determining the highest weight reveals that for the complexification $\mathcal{H}_{S^m}^k$ of $\mathcal{H}_{S^m}^k(\mathbf{R})$, we have

$$(4.3) \qquad \mathcal{H}_{S^m}^k \cong V_{SO(m+1)}^{(k,0,\ldots,0)}.$$

In general, the dimension of $V_{SO(m+1)}^\rho$ should (and in fact can) be determined in terms of ρ and m. This is given the Weyl dimension formula which, for $m+1 = 2\ell+1$, reads as

$$(4.4) \qquad \begin{aligned} \dim_{\mathbf{C}} V_{SO(m+1)}^\rho &= \frac{\prod_{r<s}(\rho_r - \rho_s - r + s)(\rho_r + \rho_s + 2\ell+1-r-s)}{\prod_{r<s}(-r+s)(2\ell+1-r-s)} \\ &\times \frac{\prod_{t=1}^\ell (\rho_t + \ell + 1/2 - t)}{\prod_{t=1}^\ell (\ell + 1/2 - t)} \end{aligned}$$

and, for $m + 1 = 2\ell$, is

(4.5) $\qquad \dim_{\mathbf{C}} V^{\rho}_{SO(m+1)} = \dfrac{\prod_{r<s}(\rho_r - \rho_s - r + s)(\rho_r + \rho_s + 2\ell - r - s)}{\prod_{r<s}(-r+s)(2\ell - r - s)}.$

PROBLEM: Using (4.3) verify (3.17).

The Branching Theorem which follows (and which was promised above) gives a precise account on the decomposition of $V^{\rho}_{SO(m+1)}|_{SO(m)}$ into irreducible complex $SO(m)$-modules.

THEOREM 4.1. *As $SO(m)$-modules, we have*

$$V^{\rho}_{SO(m+1)}|_{SO(m)} \cong \sum_{\sigma} V^{\sigma}_{SO(m)},$$

where the summation runs over all $\sigma \in \mathbf{Z}^{|[m/2]|}$ for which

$$\rho_1 \geq \sigma_1 \geq \ldots \geq \rho_\ell \geq |\sigma_\ell| \ \text{if} \ m + 1 = 2\ell + 1$$

and

$$\rho_1 \geq \sigma_1 \geq \ldots \geq \sigma_{\ell-1} \geq |\rho_\ell| \ \text{if} \ m + 1 = 2\ell. \sqrt{}$$

(For a proof, see [Börner] or [Zhelobenko].)

REMARK: Notice that (3.19) follows easily from Theorem 4.1 as a special case.

We now turn to the main point, i.e. the decomposition of $S^2(\mathcal{H}^k_{S^m})$ into irreducible components.

First note that, for $m = 2$, *all* irreducible complex $SO(3)$-modules are class 1 with respct to the pair $(SO(3), SO(2))$ (and hence they are of the form $\mathcal{H}^k_{S^2}$). This follows either by looking at the possible highest weights (that are plain integers) and using (4.3) or by verifying the existence of a nonzero vector left fixed by $SO(2)$ in an irreducible complex $SO(3)$-module. Hence, by Theorem 3.6, the $SO(3)$-module $\mathcal{E}_{\lambda_k}$ associated to a standard minimal immersion $f_{\lambda_k} : S^2 \to S^{2k}$ is trivial. We obtain what is known as the Rigidity Theorem of [Calabi]:

89

THEOREM 4.2. *Any full λ_k-eigenmap $f : S^2 \to S^n$ is standard, i.e. $n = 2k$ and the components of f comprise an orthonormal basis in $\mathcal{H}^k_{S^2}(\mathbf{R})$.* $\checkmark$

REMARK: This result was originally formulated as rigidity for minimal immersions. That version clearly follows since $\mathcal{M}_{\lambda_k} \subset \mathcal{L}_{\lambda_k}$ (cf. Theorem 2.7 of Chapter I). Henceforth we can (and will) assume that $m \geq 3$ and $k \geq 2$.

THEOREM 4.3. *For $m = 3$:*

$$(4.6) \qquad S^2(\mathcal{H}^k_{S^3}) \cong V^{(2k,0)}_{SO(4)} \oplus \sum_{j=1}^{|[k/2]|} \{ V^{(2k-2j,2j)}_{SO(4)} \oplus V^{(2k-2j,-2j)}_{SO(4)} \} \oplus S^2(\mathcal{H}^{k-1}_{S^3}),$$

and, for $m \geq 4$:

$$(4.7) \qquad S^2(\mathcal{H}^k_{S^m}) \cong \sum_{j=0}^{|[k/2]|} V^{(2k-2j,2j,0,\dots,0)}_{SO(m+1)} \oplus S^2(\mathcal{H}^{k-1}_{S^m}).$$

The entire final section of this chapter is devoted to the proof of this theorem. Here we constrain ourselves to derive various consequences of this important result. First, the long-waited decomposition of the $SO(m+1)$-module $\mathcal{E}_{\lambda_k} \otimes_{\mathbf{R}} \mathbf{C}$:

THEOREM 4.4. *Let $m \geq 3$ and $k \geq 2$ and denote by $\mathcal{E}_{\lambda_k}$ the $SO(m+1)$-module associated to a standard minimal immersion $f_{\lambda_k} : S^m \to S^{n(k)}$. Then, for $m = 3$, we have*

$$(4.8) \qquad \mathcal{E}_{\lambda_k} \otimes_{\mathbf{R}} \mathbf{C} \cong S^2(\mathcal{H}^k_{S^3})/\{\sum_{j=0}^k \mathcal{H}^{2j}_{S^3}\} \cong \sum_{\substack{(a,b)\in\triangle \\ a,b \text{ even}}} (V^{(a,b)}_{SO(4)} \oplus V^{(a,-b)}_{SO(4)}),$$

and, for $m \geq 4$, we have

$$(4.9) \qquad \mathcal{E}_{\lambda_k} \otimes_{\mathbf{R}} \mathbf{C} \cong S^2(\mathcal{H}^k_{S^m})/\{\sum_{j=0}^k \mathcal{H}^{2j}_{S^m}\} \cong \sum_{\substack{(a,b)\in\triangle \\ a,b \text{ even}}} V^{(a,b,0,\dots,0)}_{SO(m+1)},$$

where $\triangle \subset \mathbf{R}^2$ denotes the closed triangular domain with vertices $(2,2)$, (k,k) and $(2k-2,2)$.

PROOF: Every complex class 1 module with respect to the pair $(SO(m+1), SO(m))$ is of the form $V^{(a,0,\dots,0)}_{SO(m+1)} \cong \mathcal{H}^a_{S^m}$, $a \geq 0$. This is a direct consequence of Theorem

4.1. (Notice that here we could have avoided the use of the Branching Rule. In fact, given a *real* class 1 module V with respect to $(SO(m+1), SO(m))$, by the construction before Theorem 2.2, Φ_V maps V isomorphically onto an irreducible component of $C^\infty(S^m)$. Now, it is elementary that $\sum_{a=0}^{\infty} \mathcal{H}_{S^m}^a(\mathbf{R})$ is L^2-dense in $C^\infty(S^m)$ so that $V \cong \mathcal{H}_{S^m}^a(\mathbf{R})$ for some $a \geq 0$. Complexifying and using the remark before Theorem 2.2, the claim follows.) Now, by Theorem 3.6, we only have to discard the class 1 modules with respect to $(SO(m+1), SO(m))$ from the right hand sides of (4.6) and (4.7) to obtain (4.8) and (4.9), respectively. $\checkmark$

The first immediate result that follows from the decompositions (4.8) and (4.9), is the *exact* dimension of the moduli space $\mathcal{L}_{\lambda_k}$.

THEOREM 4.5. *Let* $m \geq 3$ *and* $k \geq 2$ *and denote* $\mathcal{L}_{\lambda_k}(= \mathcal{L}_{f_{\lambda_k}})$ *the moduli space associated to a standard minimal immersion* $f_{\lambda_k} : S^m \to S^{n(\lambda_k)}$. *Then, we have*

$$(4.10) \qquad \dim \mathcal{L}_{\lambda_k} = \frac{1}{2}(n(\lambda_k)+1)(n(\lambda_k)+2) - \sum_{j=0}^{k}(n(\lambda_{2j})+1),$$

where $n(\lambda_j)$ *is given in (3.17).* $\checkmark$

REMARK: Notice that substituting (4.10) for $\dim \mathcal{L}_{\lambda_k} = \dim \mathcal{E}_{\lambda_k}$ into (2.12) of Chapter I, we obtain a lower bound for the nullity of *any* full λ_k-eigenmap $f :$ $S^m \to S^{n(\lambda_k)}$.

Next we derive a consequence of Theorem 3.2 of Chapter I on the number of geometrically distinct λ_k-eigenmaps.

THEOREM 4.6. *Let* $m \geq 3$ *and* $k \geq 3$ *and assume that*

$$(4.11) \quad (n(\lambda_k)+1)(n(\lambda_k)+2) > 2\sum_{j=0}^{k}(n(\lambda_{2j})+1) + (n(\lambda_k)+1)(m(m+1)+2).$$

Then there exist $\aleph_1$ *geometrically distinct rigid full* λ_k-*eigenmaps* $f : S^m \to S^n$. *Moreover, for each* $m \geq 3$, *there exists* $k(m) \geq 3$ *such that (4.11) is satisfied for all* $k \geq k(m)$.

PROOF: (4.11) is the translation of (3.4) of Chapter I using (4.10) above. For fixed $m \geq 3$, the left hand side of (4.11) is a polynomial of degree $2(m-1)$ in k. However, on the right hand side, the first term is a polynomial of degree m in k while the second is of degree $m - 1$. $\checkmark$

We now turn to the question of uniqueness of standard minimal immersions. The following result implies, in particular, that the only equivariant λ_k-eigenmaps are the standard minimal immersions.

THEOREM 4.7. Let $f : S^m \to S^n$ be a full λ_k-eigenmap such that f is equivariant with respect to a homomorphism $\rho_f : SO(m) \to SO(n + 1)$, where $SO(m)$ acts on S^m via the inclusion $SO(m) \subset SO(m + 1)$. Then f is a standard minimal immersion.

PROOF: The isotropy subgroup of $\langle f \rangle_{f_{\lambda_k}} \in \mathcal{L}_{\lambda_k} \subset \mathcal{E}_{\lambda_k}$ contains $SO(m)$ or, in other words, $\langle f \rangle_{f_{\lambda_k}}$ is left fixed by $SO(m)$. On the other hand, by Theorem 3.6, no component of $\mathcal{E}_{\lambda_k}$ is class 1 with respect to the pair $(SO(m + 1), SO(m))$. Thus $\langle f \rangle_{f_{\lambda_k}} = 0$ and so f is standard. $\checkmark$

PROBLEM: Sow that, for $k \geq 2$, we have

$$\dim \mathrm{Fix}\,(SO(m - 1), \mathcal{E}_{\lambda_k}) = \begin{cases} \dfrac{k(k^2 - 1)}{3}, & \text{if } m = 3, \\[3mm] \dfrac{k(k^2 - 1)}{6}, & \text{if } m \geq 4. \end{cases}$$

Describe the boundary of $\mathrm{Fix}\,(SO(2), \mathcal{L}_{\lambda_2}) = \mathrm{Fix}\,(SO(2), \mathcal{E}_{\lambda_2}) \cap \mathcal{L}_{\lambda_2}$ explicitly. (Hint: Use Theorem 4.1 twice.)

PROBLEM**: $\mathrm{Fix}\,(T, \mathcal{L}_{\lambda_k})$ parametrizes those full λ_k-eigenmaps $f : S^m \to S^n$ that are equivariant with respect to a homomorphism $\rho_f : T \to SO(n + 1)$, where $T \subset SO(m + 1)$ is the standard maximal torus. Prove that, for $m = 3$ and $k = 2$, $\mathrm{Fix}\,(T, \mathcal{E}_{\lambda_k})$ is a plane spanned by $\langle H \rangle_{f_{\lambda_2}}$ and $\langle H' \rangle_{f_{\lambda_2}}$ (cf. (4.5) and (6.13) of

Chapter I). Moreover, for $m = 4$ and $k \geq 2$, show that

$$(4.12) \qquad \dim \operatorname{Fix}(T, \mathcal{E}_{\lambda_k}) = \begin{cases} \dfrac{k(k+2)(k^2 + 4k - 3)}{24}, & \text{if } k \text{ is even,} \\[4mm] \dfrac{(k^2 - 1)(k^2 + 6k + 6)}{24}, & \text{if } k \text{ is odd,} \end{cases}$$

(Hint: Use Kostant's formula for the multiplicity of zero as a weight in each of the components of $\mathcal{E}_{\lambda_k} \otimes_{\mathbf{R}} \mathbf{C}$ given in (4.8) and (4.9) (cf. [Humphreys]). Warning: (4.12) is a massive combinatorial computation.)

We are now ready to accomplish the classification of all full λ_2-eigenmaps f : $S^3 \to S^n$, $2 \leq n \leq 8$, $n \neq 3$ (cf. Theorem 4.1 of Chapter I). An abundance of examples have already been given throughout Chapter I. In view of Theorems 4.2 and 4.4, the moduli space $\mathcal{L}_{\lambda_2} \subset \mathcal{E}_{\lambda_2} (\subset S^2(\mathcal{H}_{S^3}^2(\mathbf{R})), \mathcal{H}_{S^3}^2(\mathbf{R}) \cong \mathbf{R}^9)$ of such maps corresponds to the lowest nonrigid (=nontrivial) case. As we shall see in a moment, in contrast with Theorem 4.6, the natural saturation on $\mathcal{L}_{\lambda_2}$ is finite modulo the action of $O(4)$ (in fact, there is a single cell modulo $O(4)$ in every admissibile range dimension $2 \leq n \leq 8$, $n \neq 3$.)

For the sake of formulating the classification theorem we unify the notations and introduce $f_n : S^3 \to S^n$, $2 \leq n \leq 8$, $n \neq 3$, to be the full λ_2-eigenmap given as:

$$\begin{aligned} f_2 &= H \text{ (cf. (4.3) and (4.5))} \\ f_4 &= f_\otimes \cong H \oplus H' \text{ (cf. (6.12) and Example 7.2)} \\ f_5 &= V \text{ (cf. (4.12))} \\ (4.13) \qquad f_6 &= H' \oplus V \text{ (cf. Example 7.3)} \\ f_7 &= V \oplus V' \text{ (cf. Example 7.4)} \\ f_8 &= f_{\lambda_2} \text{ (cf. (4.6)),} \end{aligned}$$

where all references are in Chapter I.

THEOREM 4.8. *Full λ_2-eigenmaps of S^3 into S^n exist iff $2 \leq n \leq 8$ and $n \neq 3$. Moreover, if $f : S^3 \rightarrow S^n$ is a full λ_2-eigenmap then there exists a $a \in O(4)$, $U \in O(n+1)$ and a symmetric positive definite matrix $A \in S^2(\mathbf{R}^{n+1})$ such that*

$$(4.14) \qquad\qquad U \circ f \circ a = A \circ f_n,$$

where $f_n : S^3 \rightarrow S^n$ is given in (4.13). Equivalently, the moduli space $\mathcal{L}_{\lambda_2}$ modulo the action of $O(4)$ is saturated by the open cells I_{f_n}, $2 \leq n \leq 8$, $n \neq 3$. Finally, I_{f_2} = point, I_{f_4} = segment, I_{f_5} = 2 − disk, I_{f_6} = (finite) solid cone, $\dim I_{f_7}$ = 5 and $\dim I_{f_8} = 10$.

PROOF: The first statement is clear from the cast (4.13) and Theorem 4.1 of Chapter I. The last follows from various results and examples of Chapter I with exact references in (4.13) with the exception of $\dim I_{f_8} = 10$ which will be shown below. We will prove (4.14) by describing the topology of the natural saturation $\mathcal{I}_{\lambda_2}$ on $\mathcal{L}_{\lambda_2}$ cell by cell.

Recall from the proof of Theorem 4.1 of Chapter I that the $O(4)$-orbit of $\langle H \rangle_{f_{\lambda_2}}$ consists of two copies $\mathbf{R}P_1^2$ and $\mathbf{R}P_2^2$ of the real projective plane imbedded in the boundary of $\mathcal{L}_{\lambda_2}$. For definiteness, put $SO(4)\langle H \rangle_{f_{\lambda_2}} = \mathbf{R}P_1^2$ and $SO(4)\langle H' \rangle_{f_{\lambda_2}} = \mathbf{R}P_2^2$. Let V_1 denote the *affine* span of $\mathbf{R}P_1^2$ and $\mathbf{R}P_2^2$ in $\mathcal{E}_{\lambda_2}$, respectively. Since $SO(4)$ acts on $\mathcal{E}_{\lambda_2}$ with no (nonzero) fixed points (Theorem 4.7), V_1 and V_2 are actually $SO(4)$-submodules of $\mathcal{E}_{\lambda_2}$. We first claim that

$$(4.15) \qquad\qquad \mathcal{E}_{\lambda_2} = V_1 \oplus V_2$$

is an orthogonal direct sum with V_1 and V_2 irreducible $SO(4)$-submodules of $\mathcal{E}_{\lambda_2}$. Indeed, inspection of the Hopf map and its dual shows that the identity component of $SO(4)_H \cap SO(4)_{H'}$, is nothing but the standard maximal torus $T \subset SO(4)$. Since $SO(4)_H = U(2)$ is 4-dimensional (cf. §4 of Chapter I) this can be rephrased by saying that $SO(4)_H$ (while leaves $\langle H \rangle_{f_{\lambda_2}}$ fixed) acts transitively on $\mathbf{R}P_2^2$. We obtain that V_1 and V_2 are orthogonal in $\mathcal{E}_{\lambda_2}$. On the other hand the general decomposition (4.8) specializes here to

$$\mathcal{E}_{\lambda_2} \otimes_{\mathbf{R}} \mathbf{C} \cong V_{SO(4)}^{(2,2)} \oplus V_{SO(4)}^{(2,-2)}$$

yielding that $\mathcal{E}_{\lambda_2}$ has (at most) two irreducible components. (4.15) along with the irreducibility of V_1 and V_2 follow. In addition, the Weyl dimension formula (4.5) gives

$$\dim V_1 = \dim V_2 = \dim_{\mathbf{C}} V_{SO(4)}^{(2,2)} = \dim_{\mathbf{C}} V_{SO(4)}^{(2,-2)} = 5$$

and we pick up the remaining $\dim I_{f_s} = 10$ of the last statement. To understand $\mathcal{L}_{\lambda_2} \subset \mathcal{E}_{\lambda_2}$, we have to describe the action of $SO(4)$ on $\partial \mathcal{L}_1 = \partial \mathcal{L}_{\lambda_2} \cap V_1$ or, what is equivariantly homemorphic, the unit 4-dimensional sphere S_{V_1}. We claim that the $SO(4)$-orbits on S_{V_1} form a homogeneous (isoparametric) family of hypersurfaces of degree 3 with two (necessarily) antipodal singular orbits corresponding to the focal varieties that are minimally imbedded real projective planes (as Veronese surfaces). This is fairly clear since the only possible dimensions of the $SO(4)$-orbits on S_{V_1} are 2 and 3 and the latter must occur (since otherwise S_{V_1} would split topologically). Since (a homothetic copy of) $\mathbf{R}P_1^2$ is already in S_{V_1} the homogeneous family of hypersurfaces ($SO(4)$-orbits) must be of degree 3 and the claim follows. Passing to $\partial \mathcal{L}_1$ we see what was promised in the remark after Theorem 4.1 of Chapter I, namely, that $\mathbf{R}P_1^2$ lies in $\partial \mathcal{L}_1 \subset V_1$ as a minimally imbedded Veronese surface.

What is the antipodal of $\mathbf{R}P_1^2$ in $\partial \mathcal{L}_1$? Computation or inspection shows that $\langle V \rangle_{f_{\lambda_2}} \in \mathbf{R} \cdot \langle H \rangle_{f_{\lambda_2}}$, where $V : S^3 \to S^5$ is the complex Veronese map defined in (4.12) of Chapter I. By range dimension, $\langle V \rangle_{f_{\lambda_2}} \in \partial \mathcal{L}_{\lambda_2}$ so that the antipodal of $\mathbf{R}P_1^2$ is $SO(4) \cdot \langle V \rangle_{f_{\lambda_2}}$.

In the end of §4 of Chapter I we showed that $\bar{I}_V (= \iota(\mathcal{L}_V)$; cf. Proposition 2.5 of Chapter I) is a 2-dimensional disk with boundary

$$\partial I_V = \{\langle H_\alpha \rangle_{f_{\lambda_2}} | \alpha \in \mathbf{R}\},$$

where H_α is given in (4.13) of Chapter I. As ∂I_V has no place else to belong, we have $\partial I_V \subset \mathbf{R}P_1^2$. Moreover, the (diagonal) center $S^1 \subset U(2)(\subset SO(4))$ rotates ∂I_V by (4.14) of Chapter I. Finally, $SO(4)_V = SO(4)_H = U(2)$ so that away from $\langle V \rangle$, at each point of I_V the isotropy subgroup is 3-dimensional. Restricting to a radial segment of I_V, for reasons of dimension, we obtain that

$$SO(4) \cdot \bar{I}_V = \partial \mathcal{L}_1.$$

Equivalently, modulo $SO(4)$, the natural saturation on $\partial \mathcal{L}_1$ contains exactly two open cells: I_H and I_V.

The same description applies to $\partial \mathcal{L}_2 = \partial \mathcal{L}_{\lambda_2} \cap V_2$ since $H' = H \circ \mathrm{diag}\,(1,1,-1,1)$. We obtain that

$$SO(4) \cdot \bar{I}_{V'} = \partial \mathcal{L}_2$$

We now consider $f_\otimes : S^3 \to S^4$. By Example 7.2 of Chapter I, $\langle f_\otimes \rangle_{f_{\lambda_2}} = \langle H \oplus H' \rangle_{f_{\lambda_2}}$ and $\bar{I}_{f_\otimes}$ is a segment connecting $\langle H \rangle_{f_{\lambda_2}} \in \mathbf{R}P_1^2$ and $\langle H' \rangle_{f_{\lambda_2}} \in \mathbf{R}P_2^2$. Since every isotropy subgroup of $\mathbf{R}P_1^2$ acts transitively on $\mathbf{R}P_2^2$, the group $SO(4)$ acts transitively on the set of all segments joining $\mathbf{R}P_1^2$ and $\mathbf{R}P_2^2$ whose union is therefore equal to $SO(4)(\bar{I}_{f_\otimes}) \subset \partial \mathcal{L}_{\lambda_2}$.

Next according to Example 7.3, $\bar{I}_{H' \oplus V}$ is a (finite) solid cone with base 2-disk $\bar{I}_V \subset \partial \mathcal{L}_1$ and vertex $\langle H' \rangle_{f_{\lambda_2}} \in \mathbf{R}P_2^2$. By birth, $\langle H' \oplus V \rangle_{f_{\lambda_2}}$ is on the center segment joining $\langle H' \rangle_{f_{\lambda_2}}$ and $\langle V \rangle_{f_{\lambda_2}}$. Moreover, the center $S^1 \subset U(2)$ rotates the cone $\bar{I}_{H' \oplus V}$. Since $\langle V \rangle_{f_{\lambda_2}} \in \mathbf{R} \cdot \langle H \rangle_{f_{\lambda_2}}$, it follows as above that $SO(4)$ acts transitively on the set of cones with base 2-disk a 2-cell on $\partial \mathcal{L}_1$ and vertex in $\mathbf{R}P_2^2$. The union of these cones then coincides with $SO(4) \cdot \bar{I}_{H' \oplus V}$. The same conclusion holds for V_1 and V_2 interchanged (by applying $\mathrm{diag}\,(1,1,-1,1) \in O(4)$). Finally, according to Example 7.4 of Chapter I, $\bar{I}_{V \oplus V'}$ is the (5-dimensional) convex hull of $\bar{I}_V$ and $\bar{I}_{V'}$. $SO(4)$ acts again transitively on the set of convex hulls spanned by *any* pairs of 2-cells (I_1, I_2) with $I_j \in \partial \mathcal{L}_j$, $j = 1,2$. The union of this set is contained in $SO(4)(\bar{I}_{V \oplus V'}) \subset \partial \mathcal{L}_{\lambda_2}$.

Summarizing, we obtain that $O(4)(\bar{I}_{H'\oplus V} \cup \bar{I}_{V\oplus V'}) \subset \partial \mathcal{L}_{\lambda_2}$ contains all segments that join $\partial \mathcal{L}_1$ and $\partial \mathcal{L}_2$ and hence it coincides with the whole $\partial \mathcal{L}_{\lambda_2}$. Since

$$O(4)(\bar{I}_{H'\oplus V} \cup \bar{I}_{V\oplus V'}) = O(4)(\cup_{\substack{n=2 \\ n\neq 3}}^{7} I_{f_n})$$

the proof is complete. $\checkmark$

PROBLEM: Let $C \in \mathcal{L}_{\lambda_k}$ and denote by $\lambda_{\max}$ the largest eigenvalue of C. Show that $\lambda_{\max} \geq 0$ with equality iff $C = 0$. Assuming $C \in \partial \mathcal{L}_{\lambda_k}$ prove that $o(C) = -\lambda_{\max}^{-1} C$ is in $\partial \mathcal{L}_{\lambda_k}$. Then $o(C)$ is said to be the *opposite* of C. (In particular, $o(\langle H \rangle_{f_{\lambda_k}}) = \langle V \rangle_{f_{\lambda_k}}$.) In general, if $C = \langle f \rangle_{f_{\lambda_k}}$, where $f : S^m \to S^n$ is a full λ_k-eigenmap, then the multiplicity of the largest eigenvalue of $o(C)$ is $n(\lambda_k) - n$. Determine the multiplicity of the largest eigenvalue of $\langle H \rangle_{f_{\lambda_k}}$.

PROBLEM: Let $v^0, \dots, v^m \in \mathbf{R}^n$ be *unit* vectors such that $\sum_{i=0}^{m} v^i = 0$. Show that

$$f : S^m \to S^{n+\frac{m(m+1)}{2}-1}$$

given by

$$f(x_0, \dots, x_m) = (v^0 x_0^2 + \dots + v^m x_m^2,$$
$$\sqrt{2 - 2\langle v^0, v^1 \rangle} x_0 x_1, \dots, \sqrt{2 - 2\langle v^{m-1}, v^m \rangle} x_{m-1} x_m)$$

is a well-defined λ_2-eigenmap. Using this, verify that there exist full λ_2-eigenmaps of S^m into $S^{n(\lambda_2)-p}$ for all $0 \leq p \leq m - 2$. Work out the special case when $v^0, \dots, v^m \in \mathbf{R}^2$ are vertices of a regular polygon. Show that the equilateral triangle corresponds to the Veronese map

$$f : S^2 \to S^4.$$

What is the analogue of this for $m \geq 3$?

PROBLEM: Let F be a C^∞-function on S^4 defined by

$$F(x, y, z, u, v) = u^3 - 3uv^2 + \frac{3}{2}u(x^2 + y^2 - 2z^2) + \frac{3\sqrt{3}}{2}v(x^2 - y^2) + 3\sqrt{3}xyz.$$

Show that $\operatorname{grad} F$ defines a λ_2-eigenmap $f : S^4 \to S^4$. (F is what is called an isoparametric function on S^4 of degree 3. Its level hypersurfaces define an isoparametric family of hypersurfaces (cf. [É.Cartan;2]).) Generalize the statement to the gradient of any isoparametric function on a sphere (cf. [Münzner]).

In the last part of the proof we observe that the isotropy subgroup of any map corresponding to a point on the segment joining $I_V \setminus \{V\}$ and $I_{V'} \setminus \{V'\}$ is finite. By the Principal Isotropy Theorem (cf. [Bredon]) on an open dense subset of $\mathcal{L}_{\lambda_2}$ the isotropy subgroup is finite. Since $\mathcal{E}_{\lambda_2} \subset \mathcal{E}_{\lambda_k}$, $k \geq 2$, the same is true for the moduli space $\mathcal{L}_{\lambda_k}$ associated to a standard minimal immersion $f_{\lambda_k} : S^3 \to S^{n(\lambda_k)}$. Finally, for $m \geq 4$, $SO(m)$ is simple and the irreducible $SO(m)$-modules that have nontrivial connected isotropy subgroup have been explicitly determined (by highest weight) by [Wu-Yi Hsiang]. A comparison of his list with the irreducible components of $\mathcal{E}_{\lambda_k} \otimes_{\mathbf{R}} \mathbf{C}$ in (4.8) and (4.9) gives the following:

THEOREM 4.9. Let $m \geq 3$ and $k \geq 2$. Then there exists an open dense subset of $\mathcal{L}_{\lambda_k}$ at each point of which the isotropy subgroup of the action of $SO(m+1)$ is finite. Equivalently, the symmetry group of a full λ_k-eigenmap $f : S^m \to S^n$ is generically finite. $\checkmark$

PROBLEM: Given a full λ_k-eigenmap $f : S^m \to S^n$, denote by $K(f)$ the vector space of divergencefree Jacobi fields along f. Show that

(4.16) $$so(n+1) \circ f + \mathcal{E}_f \circ f + f_*(so(m+1)) \subset K(f).$$

Show, by an explicit construction, that if equality holds in (4.16) then, for every $v \in K(f)$, there exists a 1-parameter family of full λ_k-eigenmaps $f_t : S^m \to S^n$, $|t| < \varepsilon$, such that $f_0 = f$ and $\frac{\partial f_t}{\partial t}|_{t=0} = v$.

PROBLEM**: Show that, for a full λ_2-eigenmap $f : S^3 \to S^n$, equality holds in (4.16). (Hint: Use the structure of $\mathcal{L}_{\lambda_2}$ as given in the proof of Theorem 4.8 along with Proposition 3.3 of Chapter I (cf. [Toth;2]). Show that $\dim\left(K(f)/so(n+1)\circ f\right)$

is constant on each open cell. Warning: The computation leading to

$$\dim\left(K(f_n)/so(n+1)\circ f_n\right) = \begin{cases} 2, \text{ if } n = 2 \\ 5, \text{ if } n = 4 \\ 4, \text{ if } n = 5 \\ 7, \text{ if } n = 6 \\ 9, \text{ if } n = 7. \end{cases}$$

is tedious.)

PROBLEM: Show that, for $f = f_\otimes : S^5 \to S^9$, where $\otimes : \mathbf{R}^3 \times \mathbf{R}^3 \to \mathbf{R}^9$, equality does not hold in (4.16).

To close this section we give a lower estimate for the dimension of the moduli space $\mathcal{M}_{\lambda_k} \subset \mathcal{F}_{\lambda_k}$ that parametrizes minimal immersions. This is the analogue of Theorems 4.4 and 4.5 in the use of Theorem 3.7. Notice that, by Problem* after the proof of Theorem 3.3, we can assume that $k \geq 4$.

THEOREM 4.10. Let $k \geq 4$, and denote $\mathcal{F}_{\lambda_k}$ the $SO(m+1)$-module associated to a standard immersion $f_{\lambda_k} : S^n \to S^{n(\lambda_k)}$. Then, for $m = 3$, we have

$$(4.17)\quad S^2(\mathcal{H}^k_{S^3})/\{\sum_{j=0}^{k}\mathcal{H}^{2j}_{S^3} \oplus \sum_{j=1}^{k-1}(V^{(2j,2)}_{SO(4)} \oplus V^{(2j,-2)}_{SO(4)})\} \cong \sum_{\substack{(a,b)\in\Delta^0 \\ a,b\,\text{even}}} (V^{(a,b)}_{SO(4)} \oplus V^{(a,-b)}_{SO(4)})$$

and, for $m \geq 4$, we have

$$(4.18)\qquad S^2(\mathcal{H}^k_{S^m})/\{\sum_{j=0}^{k}\mathcal{H}^{2j}_{S^m} \oplus \sum_{j=1}^{k-1}V^{(2j,2,0,\ldots,0)}_{SO(m+1)}\} \cong \sum_{\substack{(a,b)\in\Delta^0 \\ a,b\,\text{even}}} V^{(a,b,0,\ldots,0)}_{SO(m+1)}$$

as $SO(m+1)$-submodules of $\mathcal{F}_{\lambda_k} \otimes_\mathbf{R} \mathbf{C}$, where $\Delta^0 \subset \mathbf{R}^2$ is the closed trianglular domain with vertices $(4,4)$, (k,k) and $(2k-4,4)$. In particular, for $m = 3$, we have

$$(4.19)\quad \dim\mathcal{M}_{\lambda_k} \geq \frac{1}{2}(n(\lambda_k)+1)(n(\lambda_k)+2)) - \sum_{j=0}^{k}(n(\lambda_{2j})+1) - 2\sum_{j=1}^{k-1}(2j-1)(2j+3)$$

99

and, for $m \geq 4$, we have

$$
\dim \mathcal{M}_{\lambda_k} \geq \frac{1}{2}(n(\lambda_k) + 1)(n(\lambda_k) + 2) - \sum_{j=0}^{k}(n(\lambda_{2j}) + 1)
$$

(4.20)
$$
- \frac{1}{2}\frac{m+1}{m-1}\sum_{j=1}^{k-1}\frac{2j-1}{2j+1}(2j+m)(4j+m-1)\binom{2j+m-3}{2j},
$$

where $n(\lambda_k)$ is given in (3.17).

PROOF: The decomposition of $\bar{\mathcal{F}}_{\lambda_k}$ in (4.17) and (4.18) follow from what was said above. We now apply the Weyl dimension formulae (4.4)-(4.5) to obtain (4.19) and (4.20). $\checkmark$

REMARK: The first nonrigid range is for $m = 3$ and $k = 4$. In this case, (4.19) implies that $\dim \mathcal{M}_{\lambda_4} \geq 18$. [Muto] showed that actually $\dim \mathcal{M}_{\lambda_4} = 18$. It is not known whether the lower estimates for $\dim \mathcal{M}_{\lambda_k}$ in (4.19) and (4.20) are sharp.

§5. Decomposition of $\mathcal{H}^k_{S^m} \otimes \mathcal{H}^l_{S^m}$ and $S^2(\mathcal{H}^k_{S^m})$

We first summarize some basic facts about tensor representations of the special orthogonal group $SO(m + 1)$. For details, see [Weyl][Börner][Humphreys]. Let $\mathcal{P} = \mathbf{C}[x_0, \ldots, x_m]$ denote the ring of polynomials with complex coefficients in the variables $x_0, \ldots, x_m$. Actually, $\mathcal{P}$ is a graded algebra over $\mathbf{C}$ and, for $k \geq 0$, we denote by $\mathcal{P}^k \subset \mathcal{P}$ the complex linear subspace of homogeneous polynomials of degree k. For $k < 0$, we put $\mathcal{P}^k = \{0\}$. Moreover, $\mathcal{P}$ is a $GL(m+1; \mathbf{C})$-module, where the action of $a \in GL(m+1; \mathbf{C})$ on $\mu \in \mathcal{P}$ is given by

$$
a \cdot \mu = \mu \circ a^{-1}.
$$

Clearly, this preserves the grading so that we can consider $\mathcal{P}^k$ as a $GL(m+1; \mathbf{C})$-module. Though most of what will be said below applies perfectly well to this

100

module over $GL(m + 1; \mathbf{C})$, our primary concern in the sequel is to consider $\mathcal{P}^k$ as an $SO(m + 1)$-*module* by restriction. Using Weyl's terminology it is convenient to view $\mathcal{P}^k$ as an $SO(m + 1)$-submodule of the Weyl's space $\otimes^k \mathbf{C}^{m+1}$ of tensors of rank k over $\mathbf{C}^{m+1}$. Actually, $\mathcal{P}^k$ is contained in the $SO(m + 1)$-submodule of *symmetric* tensors of rank k. In what follows, we are interested in how to decompose the $SO(m + 1)$-submodule $\mathcal{P}_0^k \subset \mathcal{P}^k$ of traceless tensors of rank k into irreducible components. The symmetric group S_k on k letters acts on $\otimes^k \mathbf{C}^{m+1}$ by permuting the factors. This action commutes with the $SO(m + 1)$-module structure so that S_k also acts on $\mathcal{P}_0^k$. By extension, the group algebra $\mathbf{Z}_{S_k}$ also acts on $\mathcal{P}_0^k$. The principal result, due to Young, describes how to recognize the $SO(m+1)$-irreducible components of $\mathcal{P}_0^k$ in terms of $\mathbf{Z}_{S_k}$ acting on it. In fact, let $r_1 \geq \ldots \geq r_n \geq 0$ be integers with $r_1 + \ldots + r_n = k$ and consider the *Young diagram* Σ_r, $r = (r_1, \ldots, r_n)$:

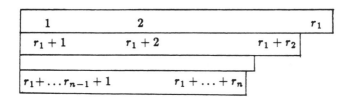

consisting of n rows with row lengths $r_1, \ldots, r_n$. Let $R(\Sigma_r)$ (resp. $C(\Sigma_r)$) denote the set of all permutations in S_k that preserve the rows (resp. columns) of Σ_r. We define the *Young symmetrizer* as the element

$$\varepsilon(\Sigma_r) = \sum_{\substack{p \in R(\Sigma_r) \\ q \in C(\Sigma_r)}} \operatorname{sgn}(q) q p$$

of the group ring $\mathbf{Z}_{S_k}$. The main result of Young (cf. [Weyl;2]) is states that, for each Young diagram Σ_r, the $SO(m+1)$-submodule $\varepsilon(\Sigma_r)\mathcal{P}_0^k$ of $\mathcal{P}_0^k$ is a multiple of an irreducible $SO(m + 1)$-submodule of $\mathcal{P}_0^k$ and all irreducible $SO(m + 1)$-submodules of $\mathcal{P}_0^k$ arise in this way. Moreover, $\varepsilon(\Sigma_r)\mathcal{P}_0^k \neq \{0\}$ iff the sum of the lengths of the first two columns of Σ_r is $\leq m + 1$. Finally, if $\Sigma_r \neq \Sigma_{r'}$ (satisfying this condition) then the corresponding irreducible $SO(m + 1)$-submodules of $\mathcal{P}_0^k$ are inequivalent.

We now turn to the decomposition of $\mathcal{H}_{Sm}^k \otimes \mathcal{H}_{Sm}^l$ and adjust the notation accordingly. We set $\mathcal{P} = \mathbf{C}[x_0, \ldots, x_m; y_0, \ldots, y_m]$ with $GL(m + 1; \mathbf{C})$ (and hence $SO(m + 1)$)-module structure given by

$$(a \cdot \mu)(x, y) = \mu(a^{-1}x, a^{-1}y), \ \mu \in \mathcal{P}, \ a \in GL(m + 1; \mathbf{C}),$$

where $x = (x_0, \ldots, x_m)$ and $y = (y_0, \ldots, y_m)$. The 'differential operators'

$$\Delta_x = \sum_{i=0}^{m} \frac{\partial^2}{\partial x_i^2}, \ \Delta_y = \sum_{i=0}^{m} \frac{\partial^2}{\partial y_i^2} \ \text{and} \ D = \sum_{i=0}^{m} \frac{\partial^2}{\partial x_i \partial y_i}$$

act on $\mathcal{P}$ as $SO(m + 1)$-module homomorphisms. Clearly, we have the commutation relations

(5.1) $$\Delta_x D = D \Delta_x \ \text{and} \ \Delta_y D = D \Delta_y.$$

Now, $\mathcal{P}$ is bigraded by the bidegree; $\mathcal{P}^{k,l} \subset \mathcal{P}$ stands for the polynomials that are homogeneous of degree k in $x_0, \ldots, x_m$ and homogeneous of degree l in $y_0, \ldots, y_m$. We put, as usual, $\mathcal{P}^{k,l} = \{0\}$ if at least one of the degrees is negative. As in the previous case, the $SO(m + 1)$-module structure preserves the bidegree so that $\mathcal{P}^{k,l}$ is an $SO(m + 1)$-module. Clearly, we have

(5.2) $$\Delta_x(\mathcal{P}^{k,l}) \subset \mathcal{P}^{k-2,l}, \ \Delta_y(\mathcal{P}^{k,l}) \subset \mathcal{P}^{k,l-2} \ \text{and} \ D(\mathcal{P}^{k,l}) \subset \mathcal{P}^{k-1,l-1}.$$

We now introduce the $SO(m + 1)$-module:

$$H^{k,l} = H_{SO(m+1)}^{k,l} = \{\mu \in \mathcal{P}^{k,l} | \Delta_x \mu = 0, \ \Delta_y \mu = 0\}.$$

Note that $H^{k,0} \cong \mathcal{H}_{Sm}^k$ and $H^{0,l} \cong \mathcal{H}_{Sm}^l$ as $SO(m + 1)$-modules and (5.2) implies

$$D(H^{k,l}) \subset H^{k-1,l-1}.$$

LEMMA 5.1. *We have*

$$H^{k,l} \cong \mathcal{H}^k_{Sm} \otimes \mathcal{H}^l_{Sm}$$

as $SO(m+1)$-modules.

PROOF: We view $\mathcal{H}^k_{Sm}$ in $\mathcal{P}^{k,0}$ and $\mathcal{H}^{0,l}_{Sm}$ in $\mathcal{P}^{0,l}$ so that $\mathcal{H}^k_{Sm} \otimes \mathcal{H}^l_{Sm}$ becomes an $SO(m+1)$-submodule of $\mathcal{P}^{k,0} \otimes \mathcal{P}^{0,l} = \mathcal{P}^{k,l}$. Choosing complementary $SO(m+1)$-submodules, we can write

$$\mathcal{P}^{k,0} = \mathcal{H}^k_{Sm} \oplus Q^{k,0} \text{ and } \mathcal{P}^{0,l} = \mathcal{H}^l_{Sm} \oplus Q^{0,l}$$

and observe that $\Delta_x | Q^{k,0}$ and $\Delta_y | Q^{0,l}$ are injective. Using these splittings, we obtain

$$\mathcal{P}^{k,l} = (\mathcal{H}^k_{Sm} \otimes \mathcal{H}^l_{Sm}) \oplus (\mathcal{H}^k_{Sm} \otimes Q^{0,l}) \oplus (Q^{k,0} \otimes \mathcal{H}^l_{Sm}) \oplus (Q^{k,0} \otimes Q^{0,l}).$$

Now, it is only the first term on the right hand side that is annihilated by both Δ_x and Δ_y. $\checkmark$

From here on, without loss of generality, we may assume that $k \geq l$. Setting

$$T^{k,l} = T^{k,l}_{SO(m+1)} = \{\mu \in H^{k,l} | D\mu = 0\}$$

we obtain the splitting

(5.3) $$H^{k,l} = T^{k,l} \oplus V^{k,l}$$

into $SO(m+1)$-submodules, where, by (5.1) and (5.2), we have

(5.4) $$V^{k,l} \subset H^{k-1,l-1}.$$

Our immediate task is to decompose $T^{k,l}$ into irreducible $SO(m+1)$-modules. To do this, we realize $\mathcal{P}^{k,l} \subset \mathcal{P}^{k+l}$ ($\subset \mathbf{C}[x_0, \ldots, x_m; y_0, \ldots, y_m]$) as an $SO(m+1)$-submodule of the Weyl's space $\otimes^{k+l} \mathbf{C}^{m+1}$. Each element $\mu \in \mathcal{P}^{k,l}$ then corresponds

103

to a tensor of rank $k + l$ (denoted by the same symbol) with components $\mu_{i_1 \ldots i_{k+l}}$ that is symmetric in $\{i_1, \ldots, i_k\}$ and $\{i_{k+1}, \ldots, i_{k+l}\}$. Moreover, $\mu \in T^{k,l}$ means that $\triangle_x \mu = 0$, $\triangle_y \mu = 0$ and $D\mu = 0$ and these translate into the single condition that contraction of μ (as a tensor) with respect to any two indices is zero. In particular, $T^{k,l} \subset \mathcal{P}_0^{k,l}$ so that Young's theory applies to $T^{k,l}$.

LEMMA 5.2. *Let Σ_r be a Young diagram corresponding to $r_1 \geq \ldots \geq r_n \geq 0$, $r_1 + \ldots + r_n = k + l$. Then*

$$(5.5) \qquad \qquad \varepsilon(\Sigma_r) T^{k,l} \neq 0$$

iff $n = 2$, i.e. $r = (r_1, r_2)$, and $r_2 \leq l$.

PROOF: Assume that $n > 2$ with $r_n > 0$. Then the first column of Σ_r contains, after any permutation in $R(\Sigma_r)$, a pair (i, j) with $1 \leq i, j \leq k$ or $k + 1 \leq i, j \leq k + l$. Using the transposition $(ij) \in C(\Sigma_r)$, we obtain

$$\sum_{q \in C(\Sigma_r)} \operatorname{sgn}(q) q \cdot p\mu = 0, \ \mu \in T^{k,l},$$

for any $p \in R(\Sigma_r)$, so that $\varepsilon(\Sigma_r) T^{k,l} = 0$.

Setting now $n \leq 2$ and $r = (r_1, r_2)$, the Young diagram, after applying an element $p \in R(\Sigma_r)$, looks like

where $i_1, \ldots, i_{r_1} \leq r_1$ and $r_1 + 1 \leq j_1, \ldots, j_{r_2} \leq r_2$. Assuming that $r_2 > l$, as $r_1 + r_2 = k + l$, we get $r_1 < k$ so that, for some $s = 1, \ldots, r_2$, we have $j_s \leq k$. Any $\mu \in T^{k,l}$ is symmetric in i_s, j_s and hence

$$\sum_{q \in C(\Sigma_r)} \operatorname{sgn}(q) q \cdot p\mu = 0$$

104

follows again. Thus (5.5) also implies $r_2 \leq l$. The converse is clear. $\checkmark$

The highest weight of the irreducible $SO(m+1)$-component in $\varepsilon(\Sigma_r)\mathcal{P}_0^{k+l}$ is known (cf. [Weyl;2]). For the Young diagrams Σ_r, $r = (r_1, r_2)$, $r_2 \leq l$, we obtain, for $m \geq 4$:

$$(5.6) \qquad T^{k,l} = \sum_{i=0}^{l} m \left[V_{SO(m+1)}^{(k+l-i,i,0,\ldots,0)} : T^{k,l} \right] V_{SO(m+1)}^{(k+l-i,i,0,\ldots,0)}$$

and, for $m = 3$:

$$(5.7) \qquad T^{k,l} = \sum_{i=0}^{l} m \left[V_{SO(4)}^{(k+l-i,\pm i)} : T^{k,l} \right] \left(V_{SO(4)}^{(k+l-i,i)} \oplus V_{SO(4)}^{(k+l-i,-i)} \right),$$

where the (positive) multiplicities will be determined in the sequel. (Recall that $k \geq l$ is assumed. Note also that, for $m = 3$, the splitting inside the sum is due to the fact that $SO(4)$ is not simple.)

The case $m = 2$ is our leading guide as to what happens for $m \geq 3$. In fact, this case can be completely settled by using the Clebsch-Gordan formula (cf. [Weyl;1]). We obtain

$$\mathcal{H}_{S^2}^k \otimes \mathcal{H}_{S^2}^l = \mathcal{H}_{S^2}^{k+l} \oplus \mathcal{H}_{S^2}^{k+l-1} \oplus \left(\mathcal{H}_{S^2}^{k-1} \otimes \mathcal{H}_{S^2}^{l-1} \right)$$

so that all multiplicities are 1 and $V_{SO(3)}^{k,l} = \mathcal{H}_{S^2}^{k-1} \otimes \mathcal{H}_{S^2}^{l-1}$ and hence

$$D(H_{SO(3)}^{k,l}) = H_{SO(3)}^{k-1,l-1}.$$

THEOREM 5.3. Let $k \geq l \geq 1$. For $m \geq 4$, we have

$$(5.8) \qquad \mathcal{H}_{S^m}^k \otimes \mathcal{H}_{S^m}^l = \sum_{i=1}^{l} V_{SO(m+1)}^{(k+l-i,i,0,\ldots,0)} \oplus \left(\mathcal{H}_{S^m}^{k-1} \otimes \mathcal{H}_{S^m}^{l-1} \right)$$

and, for $m = 3$, we have

$$(5.9) \qquad \mathcal{H}_{S^3}^k \otimes \mathcal{H}_{S^3}^l = \sum_{i=1}^{l} \left(V_{SO(4)}^{(k+l-i,i)} \oplus V_{SO(4)}^{(k+l-i,-i)} \right) \oplus \left(\mathcal{H}_{S^3}^{k-1} \otimes \mathcal{H}_{S^3}^{l-1} \right).$$

105

In particular, we have

$$(5.10) \qquad\qquad D(H^{k,l}) = H^{k-1,l-1}.$$

PROOF: We use induction by n and first consider the case $n = 3$. By elementary representation theory, the complex irreducible $SU(2)$-modules are parametrized by the dimension of the representation space; there is one (up to equivalence) in each dimension. Let W_k denote the $(k+1)$-dimensional irreducible $SU(2)$-module. (Actually, or more concretely, W_k can be thought of as the $SU(2)$-module of homogeneous polynomials of degree k in $\mathbf{C}[x_0, x_1]$ but we will not need this fact in the sequel.) W_k is an $SO(3)$-module, i.e. the module structure factors through the projection $\pi : SU(2) \to SO(3)$ (cf. §4 of Chapter I) iff k is even. Clearly, $W_{2k} \cong \mathcal{H}_{S^2}^k$.

The tensor product $W_k \otimes W_l$ is an irreducible $SU(2) \times SU(2) = Spin\,(4)$-module. Here $Spin\,(4)$ is the universal cover of $SO(4)$. Restricting to the diagonal subgroup $SU(2) \subset SU(2) \times SU(2)$, we have

$$(5.11) \qquad\qquad W_k \otimes W_k \cong \mathcal{H}_{S^3}^k$$

as $SU(2)$-modules. More generally, we use the Clebsch-Gordan formula to find

$$(5.12) \qquad\qquad W_k \otimes W_l = \sum_{i=0}^{l} W_{k+l-2i}, \ k \geq l,$$

as $SU(2)$-modules. Now, restricting everything and using (5.11)-(5.12), we compute

$$\mathcal{H}_{S^3}^k \otimes \mathcal{H}_{S^3}^l = (W_k \otimes W_l) \otimes (W_k \otimes W_l) = \left(\sum_{i=0}^{l} W_{k+l-2i}\right) \otimes \left(\sum_{j=0}^{l} W_{k+l-2j}\right)$$

$$= \sum_{i=0}^{l} \left\{ (W_{k+l} \otimes W_{k+l-2i}) \oplus (W_{k+l-2i} \otimes W_{k+l}) \right\}$$

$$\oplus \left\{ \left(\sum_{i=1}^{l} W_{k+l-2i}\right) \otimes \left(\sum_{j=1}^{l} W_{k+l-2j}\right) \right\}.$$

The last tensor product rewrites as

$$\left(\sum_{i=0}^{l-1} W_{(k-1)+(l-1)-2i}\right) \otimes \left(\sum_{j=0}^{l-1} W_{(k-1)+(l-1)-2j}\right) \cong \mathcal{H}_{S^3}^{k-1} \otimes \mathcal{H}_{S^3}^{l-1}$$

as $SU(2)$-modules. Finally, as $Spin\,(4)$ (actually, $SO(4)$)-modules, we have

$$W_{k+l} \otimes W_{k+l-2i} \cong V_{SO(4)}^{(k+l-i,i)}$$

and

$$W_{k+l-2i} \otimes W_{k+l} \cong V_{SO(4)}^{(k+l-i,-i)}$$

and so (5.9) follows. Moreover, using Lemma 5.1, we obtain (5.10) for $n = 3$ as well.

As there is an annoying difference between (5.8) and (5.9), we introduce the (slightly confusing) notation that, for $m = 3$, $V_{SO(m+1)}^{(k+l-i,i,0,\ldots,0)}$ actually means $V_{SO(4)}^{(k+l-i,i)} \oplus V_{SO(4)}^{(k+l-i,-i)}$. This we will use only in the proof. The advantage is, of course, that (5.8) now includes (5.9) as a special case.

We now perform the general step of the induction with respect to n. We first claim that

$$m\left[V_{SO(m+2)}^{(k+l-i,i,0,\ldots,0)} : \mathcal{H}_{S^{m+1}}^k \otimes \mathcal{H}_{S^{m+1}}^l\right] = 1, \ i = 0,\ldots,l.$$

This is an easy application of the induction hypothesis along with the Branching Rule (Theorem 4.1). We have

$$\mathcal{H}_{S^{m+1}}^k \otimes \mathcal{H}_{S^{m+1}}^l = \sum_{i=1}^l \left(\mathcal{H}_{S^m}^k \otimes \mathcal{H}_{S^m}^l\right)$$

(5.13) $\qquad\qquad \oplus$ terms containing $\mathcal{H}_{S^m}^u \otimes \mathcal{H}_{S^m}^v$ for $u + v < k + l$.

We now consider the $SO(m + 1)$-component $V_{SO(m+1)}^{(k+l-i,i,0,\ldots,0)}$ of $V_{SO(m+2)}^{(k+l-i,i,\ldots,0)}$ (of multiplicity 1). Using the induction hypothesis, for $u + v < k + l$, we have

(5.14) $\qquad\qquad m\left[V_{SO(m+1)}^{(k+l-i,i,0,\ldots,0)} : \mathcal{H}_{S^m}^u \otimes \mathcal{H}_{S^m}^v\right] = 0$

107

and, clearly

$$(5.15) \qquad m\left[V_{SO(m+1)}^{(k+l-i,i,0,\ldots,0)} : \sum_{i=0}^{l} \mathcal{H}_{S^m}^k \otimes \mathcal{H}_{S^m}^i\right] = 1.$$

Putting (5.13)-(5.15) together, we obtain

$$m\left[V_{SO(m+2)}^{(k+l-i,i,0,\ldots,0)} : \mathcal{H}_{S^{m+1}}^k \otimes \mathcal{H}_{S^{m+1}}^l\right]$$
$$\leq m\left[V_{SO(m+1)}^{(k+l-i,i,0,\ldots,0)} : \mathcal{H}_{S^{m+1}}^k \otimes \mathcal{H}_{S^{m+1}}^l|_{SO(m+1)}\right].$$

and the right hand side is one. The left hand side is, however, at least 1 as was established in (5.6). Hence the claim follows and (5.6) simplifies to

$$T_{SO(m+2)}^{k,l} = \sum_{i=0}^{l} V_{SO(m+2)}^{(k+l-i,i,0,\ldots,0)}.$$

By (5.3) and (5.4), we obtain

$$\mathcal{H}_{S^{m+1}}^k \otimes \mathcal{H}_{S^{m+1}}^l = \sum_{i=0}^{l} V_{SO(m+2)}^{(k+l-i,i,0,\ldots,0)} \oplus V_{SO(m+2)}^{k,l},$$

where

$$(5.16) \qquad V_{SO(m+2)}^{k,l} \subset \mathcal{H}_{S^{m+1}}^{k-1} \otimes \mathcal{H}_{S^{m+1}}^{l-1}.$$

It remains to prove that equality holds in (5.16). We verify this by showing that both sides of (5.16) have the same number of irreducible components when restricted to $SO(m+1)$. The number of irreducible $SO(m+1)$-components in $V_{SO(m+2)}^{k,l}$ can be obtained from (5.13). In fact, writing it out in details

$$\mathcal{H}_{S^{m+1}}^k \otimes \mathcal{H}_{S^{m+1}}^l|_{SO(m+1)} = \sum_{i=0}^{l}(\mathcal{H}_{S^m}^k \otimes \mathcal{H}_{S^m}^i) \oplus \sum_{i=0}^{l}(\mathcal{H}_{S^m}^i \otimes \mathcal{H}_{S^m}^l)$$

$$(5.17) \qquad \oplus \sum_{i=l+1}^{k-1}(\mathcal{H}_{S^m}^i \otimes \mathcal{H}_{S^m}^l) \oplus \sum_{u=0}^{k-1}\sum_{v=0}^{l-1}(\mathcal{H}_{S^m}^u \otimes \mathcal{H}_{S^m}^v)$$

108

and using the induction hypothesis, it follows that the number of components of the first three summands is

$$\sum_{i=0}^{l}(i+2)(i+1) + \frac{(k-l-1)(l+2)(l+1)}{2} = (k+1)(l+1) + \frac{l(l+1)(3k-l+1)}{6}.$$

By branching, the number of $SO(m+1)$-components in $\sum_{i=0}^{l} V_{SO(m+2)}^{(k+l-i,i,0,\ldots,0)}$ is

$$\sum_{i=0}^{l}\sum_{u=i}^{k+l-i}(i+1) = \sum_{i=0}^{l}(k+l-2i+1)(i+1) = (k+1)(l+1) + \frac{l(l+1)(3k-l+1)}{6}.$$

Finally, again by branching

$$\mathcal{H}_{S^{m+1}}^{k-1} \otimes \mathcal{H}_{S^{m+1}}^{l-1}|_{SO(m+1)} \cong \sum_{u=0}^{k-1}\sum_{v=0}^{l-1}(\mathcal{H}_{S^m}^u \otimes \mathcal{H}_{S^m}^v)$$

and this is just the last summand in (5.17). Thus both sides of (5.16) have the same number of irreducible $SO(m+1)$-components and we are done. $\checkmark$

Finally we turn to the decomposition of $S^2(\mathcal{H}_{S^m}^k)$ and prove Theorem 4.3. We first observe that $S^2(\mathcal{H}_{S^m}^k) \subset \mathcal{H}_{S^m}^k \otimes \mathcal{H}_{S^m}^k \cong H^{k,k} \subset \mathcal{P}^{k,k}$ corresponds to the $SO(m+1)$-submodule $SH^{k,k}$ consisting of those polynomials μ for which the coefficients satisfy the condition

(5.18)
$$\mu_{i_1 \ldots i_k i_{k+1} \ldots i_{k+l}} = \mu_{i_{k+1} \ldots i_{k+l} i_1 \ldots i_k}.$$

Note also that, by (5.10), we have

$$D(SH^{k,k}) = SH^{k-1,k-1}$$

since the symmetrization interchanging $\{i_1, \ldots, i_k\}$ and $\{i_{k+1}, \ldots, i_{k+l}\}$. commutes with D. Now the $SO(m+1)$-submodule

$$ST^{k,k} = \{\mu \in SH^{k,k}|D\mu = 0\} \subset \mathcal{P}_0^{2k} \subset \otimes^{2k}\mathbf{C}^{m+1}$$

can be looked upon as the collection of all traceless tensors μ that are zero under pairwise contractions, symmetric in $\{i_1, \ldots, i_k\}$, $\{i_{k+1}, \ldots, i_{k+l}\}$ and satisfy (5.18). Using these remarks the decomposition formulae (5.8) and (5.9) will imply (4.6) and (4.7) once we establish the following:

LEMMA 5.4. Let Σ_r, $r = (r_1, r_2)$, be a Young diagram satisfying $r_1 \geq r_2 \geq 0$, $r_1 + r_2 = 2k$ and $r_2 \leq k$. Then

$$\varepsilon(\Sigma_r)ST^{k,k} \neq 0$$

iff r_2 is even.

PROOF: Let $\mu \in ST^{k,k}$ with $\varepsilon(\Sigma_r)\mu = \mu'$. Using the symmetry properties above, we compute

$$
\begin{aligned}
\mu'_{i_1...i_{r_1} j_1...j_{r_2}} &= (-1)^{r_2} \mu'_{j_1...j_{r_2} i_{r_2+1}...i_{r_1} i_1...i_{r_2}} \\
&= (-1)^{r_2} \mu'_{i_{k+1}...i_{r_1} i_1...i_{r_2} j_1...j_{r_2} i_{r_2+1}...i_k} \\
&= (-1)^{r_2} \mu'_{i_{k+1}...i_{r_1} i_1...i_{r_2} i_{r_2+1}...i_k j_1...j_{r_2}} \\
&= (-1)^{r_2} \mu'_{i_1...i_{r_1} j_1...j_{r_2}}.
\end{aligned}
$$

Thus, for r_2 odd, we obtain $\mu = 0$. The converse follows by reversing the steps above. $\sqrt{}$

CHAPTER III

COMPLEX EIGENMAPS AND MINIMAL IMMERSIONS

BETWEEN COMPLEX PROJECTIVE SPACES

§1. Construction of the moduli spaces

In §4 of Chapter I we introduced the complex vector space $\mathcal{H}^{p,q}_{\mathbf{C}P^m}$ of harmonic homogeneous polynomials of bidegree (p,q) as an irreducible $U(m+1)$-submodule of the restriction $\mathcal{H}^{p+q}_{S^{2m+1}}|_{U(m+1)}$. As eigenmaps between spheres were manufactured from real spherical harmonics, here, turning to complex, we wish to study those eigenmaps that have components in a fixed $\mathcal{H}^{p,q}_{\mathbf{C}P^m}$. Given $p > q \geq 0$, we call $f : S^{2m+1} \to S^{2n+1}$ a *(complex) eigenmap of bidegree (p,q)* if the components of f (with respect to the standard (and hence any) basis) in $\mathbf{C}^{m+1}$ ($\supset S^{2m+1}$) belong to $\mathcal{H}^{p,q}_{\mathbf{C}P^m}$. By (3.20) of Chapter II, f is then a λ_{p+q}-eigenmap, in particular, it is harmonic.

By birth, an eigenmap $f : S^{2m+1} \to S^{2n+1}$ of bidegree (p,q) is equivariant with respect to the homomorphism

$$\rho_{p-q} : S^1 \to S^1$$

between the centers of $U(m+1)$ and $U(n+1)$, where ρ_{p-q} is given by

$$\rho_{p-q}(\mathrm{diag}\,(e^{i\theta},\ldots,e^{i\theta})) = \mathrm{diag}\,(e^{i(p-q)\theta},\ldots,e^{i(p-q)\theta}), \ \theta \in \mathbf{R}.$$

This is just the translation of the fact that the center $S^1 \subset U(m+1)$ acts on $\mathcal{H}_{CP^m}^{p,q}$ by weight $p - q > 0$. In particular, f factors through the Hopf bundle maps $H : S^{2m+1} \to CP^m$ and $H : S^{2n+1} \to CP^n$ yielding a map $F : CP^m \to CP^n$ such that the diagram

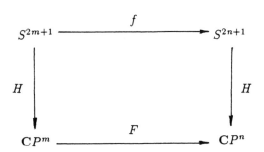

commutes. More plainly, using homogeneous coordinates, if the (complex) components of f are $f^0, \ldots, f^n$ then F is given by

$$F([z_0 : \ldots : z_m]) = [f^0(z_0, \ldots, z_m) : \ldots : f^n(z_0, \ldots, z_m)], \quad [z_0 : \ldots : z_m] \in CP^m.$$

We say that $F : CP^m \to CP^n$ is *induced* by f. Note that F determines f uniquely. We now use (a particular case of) a general result known as the Smith Reduction Theorem (cf. [Smith][Eells-Lemaire]). In fact, suppose that

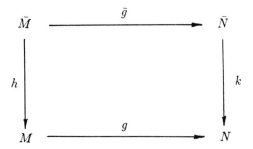

is a commutative diagram, where the vertical arrows are (harmonic) Riemannian submersions with totally geodesic fibres. If $\bar{g} : \bar{M} \to \bar{N}$ is *horizontal*, i.e. its

112

differential maps the horizontal distribution $(\ker h_*)^{\perp} \subset T(\bar{M})$ into the horizontal distribution $(\ker k_*)^{\perp} \subset T(\bar{N})$ then we have

$$\tau(g \circ h) = \tau(k \circ \bar{g}) = k_* \tau(\bar{g})$$

for the tension fields. In particular, if $\bar{g} : \bar{M} \to \bar{N}$ is horizontal and harmonic then, by Remark 3 following Proposition 1.2 of Chapter II, $g : M \to N$ is also harmonic.

PROPOSITION 1.1. *Let $f : S^{2m+1} \to S^{2n+1}$ be an eigenmap of bidegree (p, q). Then the induced map $F : \mathbf{C}P^m \to \mathbf{C}P^n$ is harmonic iff f is horizontal. In any case, F has degree $p - q$ (on second cohomology), in particular, $m \leq n$.*

PROOF: The first statement follows from what was said above.

For the second statement, we note that, by homogeneity, we have

$$F^* \gamma_{\mathbf{C}P^n} \cong \otimes^{p-q} \gamma_{\mathbf{C}P^m},$$

where γ denotes the canonical line bundle. Taking first Chern classes of both sides we get

$$F^* c_1(\gamma_{\mathbf{C}P^n}) = (p - q) c_1(\gamma_{\mathbf{C}P^m})$$

and the claim follows. Finally, the last statement is pure algebra since $H^*(\mathbf{C}P^m; \mathbf{Z})$ and $H^*(\mathbf{C}P^n; \mathbf{Z})$ are truncated polynomial rings. $\sqrt{}$

REMARK: It is also apparent that $F : \mathbf{C}P^m \to \mathbf{C}P^n$ induced by an eigenmap $f : S^{2m+1} \to S^{2n+1}$ of bidegree (p, q) is *nonholomorphic* unless $q = 0$.

The upshot of this section is to construct, for fixed $m \geq 1$ and $p > q \geq 0$, moduli spaces of eigenmaps $f : S^{2m+1} \to S^{2n+1}$ of bidegree (p, q) (for various n).

As usual, we see no harm in starting with a more general setting than necessary. The discussion that follows is the complex analogue of the construction of moduli spaces for eigenmaps between spheres presented in §2 of Chapter I. This allows us to be occasionally sketchy in the details.

113

A map $f : M \rightarrow S_V{}'$ of class C^∞ of a Riemannian manifold M into the unit sphere of a *Hermitian* vector space V is said to be *full (in complex sense)* if the image of $f : M \rightarrow V$ spans V *over* **C**.

Two maps $f_1 : M \rightarrow S_{V_1}$ and $f_2 : M \rightarrow S_{V_2}$, where V_1 and V_2 are Hermitian vector spaces, are said to be *unitary equivalent*, written as $f_1 \cong f_2$, if there exists a unitary transformation $U : V_1 \rightarrow V_2$ such that $U \circ f_1 = f_2$.

As in the real case, harmonicity and fullness are preserved by unitary equivalence which is then an equivalence relation on the set of all full harmonic maps $f : M \rightarrow S_V$ of a fixed Riemannian manifold M into the unit sphere S_V of a Hermitian vector space V, for various V.

Let V and V' be Hermitian vector spaces and $f : M \rightarrow S_V$ and $f' : M \rightarrow S_{V'}$ full harmonic maps. f' is said to be *derived* from f, written as $f' \leftarrow f$, if there exists a *complex* linear map $A : V \rightarrow V'$ such that $A \circ f = f'$. Clearly, A is uniquely determined by f and f' and is surjective. Also, if $f : S^{2m+1} \rightarrow S^{2n+1}$ is a full eigenmap of bidegree (p, q) then so is any full harmonic map $f' : S^{2m+1} \rightarrow S^{2n'+1}$ derived from f.

As in the real case, our first objective is to give a parametrization of the set of unitary equivalence classes of full harmonic maps that are derived from a given full harmonic map $f : M \rightarrow S_V$, where V is a fixed Hermitian vector space.

Let $H^2 V$ denote the Euclidean vector space of Hermitian symmetric endomorphisms of V. Then $H^2 V$ is a real form of $\mathrm{Hom}\,(V, V) = V \otimes V^*$. The scalar product on $H^2(V)$ is the restriction of the Hermitian scalar product on $V \otimes V^*$ given by

$$\langle C, C' \rangle = \mathrm{trace}\,(C'^* \cdot C), \quad C, C' \in V \otimes V^*,$$

where $*$ stands for Hermitian adjoint. Note that this scalar product is automatically real valued on $H^2 V$. For a nonzero vector $v \in V$, we denote by $\mathrm{proj}\,[v] \in H^2 V$ the orthogonal projection of V onto $\mathbf{C} \cdot v$. For $v \in S_V$, we then have

$\text{proj}\,[v](w) = \langle w, v \rangle v, \; w \in V$. As Hermitian symmetric endomorphisms are diagonalizable, $H^2 V = \text{span}\,\{\,\text{proj}\,[v] | v \in S_V\}$.

Now let $f : M \to S_V$ be a full harmonic map. We put

$$\mathcal{W}_f = \text{span}\,\{\,\text{proj}\,[f(x)] | x \in M\} \subset H^2 V,$$

and

$$\mathcal{E}_f = \mathcal{W}_f^\perp \subset H^2 V.$$

Finally, let

$$\mathcal{L}_f = \{C \in \mathcal{E}_f | C + I \geq 0\}.$$

Notice that $\mathcal{L}_f$ is a convex body in $\mathcal{E}_f$. The next classification theorem can be proved in much the same way as Theorems 2.1 and 2.3 of Chapter I.

THEOREM 1.2. *Given a full harmonic map $f : M \to S_V$ of a Riemannian manifold M into the unit sphere S_V of a Hermitian vector space V, the set of unitary equivalence classes of full harmonic maps $f : M \to S_{V'}$ that are derived from f can be parametrized by $\mathcal{L}_f$. The parametrization is given by associating to the equivalence class of f' the Hermitian symmetric endomorphism*

$$\langle f' \rangle_f = A^* \cdot A - I$$

of V, where $f' = A \circ f$. If M is compact then, for $C \in \mathcal{L}_f$, we have

$$(1.1) \qquad \qquad \text{trace}\, C \leq \frac{\text{vol}\,(M)}{\lambda_{\min}} - \dim_{\mathbf{C}} V,$$

where $\text{vol}\,(M) = \int_M \nu_M$ is the volume of M and $\lambda_{\min}(> 0)$ is the smallest eigenvalue of the positive definite endomorphism

$$(1.2) \qquad \qquad Q = \int_M \text{proj}\,[f] \cdot \nu_M \in H^2 V.$$

In particular, $\mathcal{L}_f$ is compact for M compact. Equality holds in (1.1) if all eigenvalues of Q are equal.$\sqrt{}$

PROBLEM: Define and study the natural saturation $\mathcal{I}_f$ of $\mathcal{L}_f$.

Now assume that $f : M \to S_V$ is equivariant with respect to a homomorphism $\rho_f : G \to U(V)$, where G is a closed subgroup of the group of isometries of M. Then $\mathcal{E}_f \subset H^2 V \subset V \otimes V^*$ is G-submodule, where the unitary G-module structure on $V \otimes V^*$ is given by

$$a \cdot C = Ad(\rho_f(a)) \cdot C = \rho_f(a) \cdot C \cdot \rho_f(a)^*, \ C \in V \otimes V^*, \ a \in G.$$

Moreover, $\mathcal{L}_f \subset \mathcal{E}_f$ is G-invariant, in fact, for $f' \leftharpoondown f$, we have

$$a \cdot \langle f' \rangle_f = \langle f' \circ a^{-1} \rangle_f, \ a \in G.$$

PROBLEM: Derive the complex analogue of Theorem 3.2 of Chapter I about the number of geometrically distinct harmonic maps.

We now return to our special case when $M = S^{2m+1}, m \geq 1$, and $V = \mathcal{H}^{p,q}_{\mathbf{C}Pm}, p > q \geq 0$. We take as a $U(m+1)$-invariant Hermitian scalar product $\langle\,,\rangle$ on $\mathcal{H}^{p,q}_{\mathbf{C}Pm}$ the normalized L^2-scalar product

$$\langle \mu, \mu' \rangle = \frac{n(p,q) + 1}{\mathrm{vol}\,(S^{2m+1})} \int_{S^{2m+1}} \mu \bar{\mu}' \cdot \nu_{S^{2m+1}}, \ \mu, \mu' \in \mathcal{H}^{p,q}_{\mathbf{C}Pm},$$

where $\nu_{S^{2m+1}}$ is the volume form on S^{2m+1} with volume

$$\mathrm{vol}\,(S^{2m+1}) = \int_{S^{2m+1}} \nu_{S^{2m+1}} = \frac{2\pi^{m+1}}{m!}$$

(cf. [Berger]) and

$$(1.3) \quad n(p,q)+1 = \dim_{\mathbf{C}} \mathcal{H}^{p,q}_{\mathbf{C}Pm} = \binom{m+p}{p}\binom{m+q}{q} - \binom{m+p-1}{p-1}\binom{m+q-1}{q-1}.$$

$((1.3)$ can be obtained by easy inspection. It is also a particular case of the Weyl dimension formula that will be given in §3.)

116

Let $\{f^i_{p,q}\}_{i=0}^{n(p,q)} \subset \mathcal{H}^{p,q}_{\mathbf{CP}^m}$ be an orthonormal basis, which, at the same time, identifies $\mathcal{H}^{p,q}_{\mathbf{CP}^m}$ with $\mathbf{C}^{n(p,q)+1}$, and define

$$f_{p,q} = (f^0_{p,q}, \ldots, f^{n(p,q)}_{p,q}) : S^{2m+1} \to \mathbf{C}^{n(p,q)+1}.$$

Repeating the argument given in §2 of Chapter II, we obtain that $|f_{p,q}|^2 = 1$ so that

$$f_{p,q} : S^{2m+1} \to S^{2n(p,q)+1}$$

is a full eigenmap of bidegree (p,q) that is equivariant with respect to the homomorphism $\rho_{p,q} : U(m+1) \to U(n(p,q)+1)$ that defines the $U(m+1)$-module structure on $\mathcal{H}^{p,q}_{\mathbf{CP}^m} \cong \mathbf{C}^{n(p,q)+1}$. We call $f_{p,q} : S^{2m+1} \to S^{2n(p,q)+1}$ *the standard eigenmap of bidegree* (p,q). By construction, a full harmonic map $f : S^{2m+1} \to S^{2n+1}$ is an eigenmap of bidegree (p,q) iff $f \leftharpoondown f_{p,q}$. Hence $\mathcal{L}_{f_{p,q}}$ parametrizes (the unitary equivalence classes of) all eigenmaps $f : S^{2m+1} \to S^{2n+1}$ of bidegree (p,q).

It is now time to simplify the notation by setting

$$H^2_{p,q} = H^2(\mathcal{H}^{p,q}_{\mathbf{CP}^m}) \subset \mathcal{H}^{p,q}_{\mathbf{CP}^m} \otimes \mathcal{H}^{q,p}_{\mathbf{CP}^m}$$

and

$$\mathcal{W}_{p,q} = \mathcal{W}_{f_{p,q}}, \quad \mathcal{E}_{p,q} = \mathcal{E}_{f_{p,q}}, \quad \mathcal{L}_{p,q} = \mathcal{L}_{f_{p,q}}.$$

REMARK: By orthogonality, the ij-th entry of Q in (1.2) is

$$\int_{S^{2m+1}} f^i_{p,q} \bar{f}^j_{p,q} \cdot \nu_{S^{2m+1}} = \frac{\mathrm{vol}\,(S^{2m+1})}{n(p,q)+1} \delta_{ij}, \quad i,j = 0, \ldots, n(p,q),$$

so that, using Theorem 1.2, it follows that $\mathcal{E}_{p,q}$ consists of *traceless* Hermitian symmetric endomorphisms.

PROBLEM: For $m = 1$, determine the eigenmap $f_{p,q} : S^3 \to S^{2(p+q)+1}$ of bidegree (p,q) explicitly. Show that $\mathcal{L}_{p,q} = \{0\}$. (In fact, the induced map $F_{p,q} : \mathbf{CP}^1 \to \mathbf{CP}^{p+q}$ *for all* $p,q \geq 0$ comprise all harmonic maps of $\mathbf{CP}^1$ into $\mathbf{CP}^n$ (cf. [Eells-J. Wood]).) We may therefore assume from here on that $m \geq 2$.

117

EXAMPLE 1.3: For $q = 0$, $n(p,0) + 1 = \binom{m+p}{p}$ and the holomorphic map $F_{p,0}$: $CP^m \to CP^{\binom{m+p}{p}-1}$ is nothing but the *Veronese map*. Using coordinates, we have

$$f_{p,0}(z_0, \ldots, z_m) = \left(\left(\frac{p!}{i_0! \ldots i_m!}\right)^{1/2} z_0^{i_0} \ldots z_m^{i_m}\right)_{\substack{i_0+\ldots+i_m=p \\ i_0,\ldots,i_m \geq 0}}$$

Note that $V = F_{2,0} : CP^1 \to CP^2$ has been introduced in §4 of Chapter I. For all $p > 0$, we have $\mathcal{L}_{p,0} = \{0\}$. In fact, given a full eigenmap $f : S^{2m+1} \to S^{2n+1}$ of bigedree $(p,0)$, we have $f \hookleftarrow f_{p,0}$ so that $f = A \circ f_{p,0}$ holds for some $A : \mathbf{C}^{\binom{m+p}{p}} \to \mathbf{C}^{n+1}$. Writing out the condition

$$\sum_{i=0}^{n} |f^i(z)|^2 = (|z_0|^2 + \ldots + |z_m|^2)^p, \ z = (z_0, \ldots, z_m) \in \mathbf{C}^{m+1},$$

in terms of the entries of A, we obtain

$$\sum_{\substack{j,k \geq 0 \\ |j|=|\bar{k}|=p}} \left\{ \sum_{l=0}^{n} a_{lj} \bar{a}_{lk} \right\} \left(\frac{p!}{j!k!}\right)^{1/2} z^j \bar{z}^k = \sum_{\substack{j \geq 0 \\ |j|=p}} \frac{p!}{j!} z^j \bar{z}^j,$$

where we used multiindices. Comparing coefficients, it follows that A is unitary. As $\mathcal{L}_{p,0}$ is trivial for all $p > 0$, in what follows we may (and will) assume that $q > 0$.

We now turn to minimal immersions.

PROPOSITION 1.4. *For $p > q > 0$, the map $F_{p,q} : CP^m \to CP^{n(p,q)}$ induced by a full standard eigenmap $f_{p,q}$ of bidegree (p,q) is homothetic.*

PROOF: Consider the Hermitian symmetric 2-tensor

$$\omega = \sum_{j=0}^{n(p,q)} df_{p,q}^j \otimes d\bar{f}_{p,q}^j$$

on $T(S^{2m+1})$. Clearly, ω is $U(m+1)$-invariant, in particular, it projects down yielding a Hermitian symmetric 2-tensor Ω on CP^m. Denoting by $U(m) = [1] \times U(m) \subset U(m+1)$ the isotropy subgroup corresponding to $o = (1, 0, \ldots, 0) \in \mathbf{C}^{m+1}$, as $U(m)$ acts on $T_O(CP^m)(\cong \mathbf{C}^m)$, $O = H(o) = [1 : 0 : \ldots : 0] \in CP^m$, by matrix multiplication, Ω_O is a (real) constant multiple of the standard metric at O. By $U(m+1)$-invariance, this holds throughout CP^m so that $F_{p,q}$ is homothetic. $\checkmark$

We now turn to horizontality.

PROPOSITION 1.5. $f_{p,q} : S^{2m+1} \to S^{2n(p,q)+1}$ is horizontal (with respect to the Hopf maps $H : S^{2m+1} \to \mathbf{C}P^m$ and $H : S^{2n(p,q)+1} \to \mathbf{C}P^{n(p,q)}$), in particular, $F_{p,q} : \mathbf{C}P^m \to \mathbf{C}P^{n(p,q)}$ is harmonic and hence a minimal immersion.

PROOF: Using the notations of the proof of the Proposition 1.4, the isotropy representation of $U(m)$ on $T_o(S^{2m+1})$ decomposes as

$$T_o(S^{2m+1}) = \ker H_{*_o} \oplus (\ker H_{*_o})^\perp,$$

where the first term on the right hand side is the trivial $U(m)$-module and the second is isomorphic with $\mathbf{C}^m$ as a $U(m)$-module, where $U(m)$ acts on $\mathbf{C}^m$ by ordinary matrix multiplication. Schur's lemma (applied to the orthogonal projection of $(f_{p,q})_* T_o(S^{2m+1})$ onto $(f_{p,q})_*(\ker H_{*_o})^\perp$) implies that $(f_{p,q})_* \ker H_{*_o}$ and $(f_{p,q})_*(\ker H_{*_o})^\perp$ are orthogonal. $(f_{p,q})_* \ker H_*$ and $(f_{p,q})_*(\ker H_*)^\perp$ are then orthogonal everywhere on S^{2m+1} as $f_{p,q}$ is equivariant. $\sqrt{}$

We now reformulate the condition of horizontality of a full eigenmap $f : S^{2m+1} \to S^{2n+1}$ of bidegree (p,q) in terms of $\langle f \rangle_{f_{p,q}}$, where $f_{p,q} : S^{2m+1} \to S^{2n(p,q)+1}$, $p > q > 0$, is a fixed standard eigenmap. Using horizontality of $f_{p,q}$ established in Proposition 1.5, for $V_z \in \ker H_{*_z}$ and $X_z \in (\ker H_{*_z})^\perp$, we compute

$$\langle f_*(V_z), f_*(X_z) \rangle = \langle (A^*A)(f_{p,q})_*(V_z), (f_{p,q})_*(X_z) \rangle$$
$$= \langle (A^*A - I)(f_{p,q})_*(V_z), (f_{p,q})_*(X_z) \rangle$$
$$= \langle \langle f \rangle_{f_{p,q}}, (f_{p,q})_*(V_z) \cdot (f_{p,q})_*(X_z) \rangle,$$

where the dot stands for the Hermitian symmetric product. In general, given a Hermitian vector space V, for $u, v \in V$, we denote by $u \cdot v \in H^2 V$ the Hermitian symmetric endomorphism of V given by $(u \cdot v)(w) = 1/2 \langle w, u \rangle v + 1/2 \langle w, v \rangle u$, $w \in V$. Clearly, for $V = \mathcal{H}^{p,q}_{\mathbf{C}P^m}$, we have $f_{p,q}(z) \cdot f_{p,q}(z) = \operatorname{proj}[f_{p,q}(z)]$, $z \in S^{2m+1}$. We obtain that a full eigenmap $f : S^{2m+1} \to S^{2n+1}$ of bidegree (p,q) is horizontal iff $\langle f \rangle_{f_{p,q}}$ is orthogonal to the Hermitian symmetric endomorphism $(f_{p,q})_*(V_z) \cdot (f_{p,q})_*(X_z)$ of $\mathcal{H}^{p,q}_{\mathbf{C}P^m}$, for all $V_z \in \ker H_{*_z}$ and $X_z \in (\ker H_{*_z})^\perp$.

We define

$$\mathcal{A}_{p,q} = \text{span}\{(f_{p,q})_*(V_z) \cdot (f_{p,q})_*(X_z)|$$

(1.4) $$V_z \in \ker H_{*_z}, \, X_z \in (\ker H_{*_z})^\perp, \, z \in S^{2m+1}\}.$$

Clearly, $\mathcal{A}_{p,q}$ is a real $U(m+1)$-submodule of $H^2_{p,q}$. Setting

$$\mathcal{E}^*_{p,q} = \mathcal{E}_{p,q} \cap \mathcal{A}^\perp_{p,q} \subset H^2_{p,q},$$

we arrive at the following:

THEOREM 1.6. *The compact convex body*

$$\mathcal{L}^*_{p,q} = \mathcal{L}_{p,q} \cap \mathcal{A}^\perp_{p,q}$$

parametrizes the unitary equivalence classes of those full eigenmaps $f : S^{2m+1} \to$ S^{2n+1} *of bidegree* (p,q) *that are horizontal.* $\checkmark$

We now turn to minimal immersions and define the (real) $U(m+1)$-submodule

(1.5) $$\mathcal{Z}_{p,q} = \text{span}\{\text{proj}[(f_{p,q})_*X_z]|X_z \in \ker(H_{*_z})^\perp, \, z \in S^{2m+1}\},$$

where $H : S^{2m+1} \to \mathbf{C}P^m$ is the Hopf map. Assume now that $f : S^{2m+1} \to S^{2n+1}$ is a full eigenmap of bidegree (p,q) with $f = A \circ f_{p,q}$. For $X_z \in (\ker H_{*_z})^\perp \subset$ $T_z(S^{2m+1})$, $z \in S^{2m+1} \subset \mathbf{C}^{m+1}$, we compute

$$|f_*(X_z)|^2 - |(f_{p,q})_*(X_z)|^2 = |A(f_{p,q})_*(X_z)|^2 - |(f_{p,q})_*(X_z)|^2$$

$$= \langle (A^*A - I)(f_{p,q})_*(X_z), (f_{p,q})_*(X_z) \rangle$$

$$= \text{trace}(\text{proj}[(f_{p,q})_*(X_z)] \cdot \langle f \rangle_{f_{p,q}})$$

$$= \langle \langle f \rangle_{f_{p,q}}, \text{proj}[(f_{p,q})_*(X_z)] \rangle.$$

Setting

$$\mathcal{F}_{p,q} = \mathcal{Z}^\perp_{p,q} \subset H^2_{p,q},$$

and

$$\mathcal{F}^*_{p,q} = \mathcal{F}_{p,q} \cap \mathcal{A}^\perp_{p,q}$$

we arrive at the following:

120

THEOREM 1.7. *The compact convex body*

$$\mathcal{M}_{p,q} = \mathcal{F}_{p,q} \cap \mathcal{L}_{p,q} \subset \mathcal{F}_{p,q} \ (\text{resp.}\ \mathcal{M}^*_{p,q} = \mathcal{F}^*_{p,q} \cap \mathcal{L}_{p,q} \subset \mathcal{F}^*_{p,q})$$

parametrizes the unitary equivalence classes of those full eigenmaps $f : S^{2m+1} \rightarrow S^{2n+1}$ *of bidegree* (p,q) *that induce homothetic (resp. homothetic minimal) immersions* $F : \mathbf{C}P^m \rightarrow \mathbf{C}P^n$ *of the same homothety constant as that of* $F_{p,q}$. $\checkmark$

§2. $\mathcal{E}_{p,q}$ $(\mathcal{E}^*_{p,q})$ and $\mathcal{F}_{p,q}$ $(\mathcal{F}^*_{p,q})$ as $U(m+1)$-modules

We begin here with the complex version of Corollary 3.2 of Frobenius Reciprocity given in Chapter II.

PROPOSITION 2.1. *Let* V *be a finite dimensional unitary* G-*module. Given a real* K-*submodule* W_0 *of* $H^2V \subset V \otimes V^*$, *define*

$$(2.1) \qquad\qquad W = \operatorname{span}(G \cdot W_0) \subset H^2V.$$

Let $\bar{W}$ *be the sum of those complex irreducible* G-*submodules of* $V \otimes V^*$ *that when restricted to* K *contain an irreducible component of* $W_0 \otimes_{\mathbf{R}} \mathbf{C}$. *Then we have*

$$W \otimes_{\mathbf{R}} \mathbf{C} \subset \bar{W}.$$

PROOF: $H^2V \subset V \otimes V^*$ is a real form. The rest follows the same way as in the proof of Corollary 3.2 of Chapter II. $\checkmark$

Now let $f_{p,q} : S^{2m+1} \rightarrow S^{2n(p,q)+1}$ be a full eigenmap of bidegree (p,q) and choose $o = (1, 0, \ldots, 0) \in S^{2m+1}$ as a base point with isotropy subgroup $U(m) = [1] \times U(m) \subset U(m+1)$. Setting $(G, K) = (U(m+1), U(m))$, $V = \mathcal{H}^{p,q}_{\mathbf{C}P^m}$ and

121

$W_0 = \text{proj}\,[f_{p,q}(o)]$, by equivariance of $f_{p,q}$, the $U(m+1)$-submodule W in (2.1) specializes to

$$
\begin{aligned}
\mathcal{W}_{p,q} &= \text{span}\,\{\,\text{proj}\,[f_{p,q}(z)]|z \in S^{2m+1}\} \\
&= \text{span}\,\{U(m+1)\cdot(\,\text{proj}\,[f_{p,q}(o)])\} \\
&= \text{span}\,(U(m+1)\cdot W_0).
\end{aligned}
$$

COROLLARY 2.2. *Let $\bar{\mathcal{E}}_{p,q}$, denote the sum of those complex irreducible $U(m+1)$-submodules of $\mathcal{H}^{p,q}_{\mathbf{C}P^m}\otimes\mathcal{H}^{q,p}_{\mathbf{C}P^m}$ that are not class 1 with respect to $(U(m+1),U(m))$, or equivalently, do not contain $\mathcal{H}^{0,0}_{\mathbf{C}P^{m-1}}$ as a $U(m)$-submodule (by restriction). Then we have*

$$
\bar{\mathcal{E}}_{p,q} \subset \mathcal{E}_{p,q} \otimes_{\mathbf{R}} \mathbf{C}.\,\checkmark
$$

Turning to $\mathcal{Z}_{p,q} \subset H^2_{p,q}$, we first define the real linear subspace

$$
\mathcal{T} = (f_{p,q})_*(\ker H_{*_o})^{\perp} \subset \mathcal{H}^{p,q}_{\mathbf{C}P^m},
$$

where, as usual, we omitted the natural shift. Clearly, $\mathcal{T}$ is a real irreducible $U(m)$-submodule of $\mathcal{H}^{p,q}_{\mathbf{C}P^m}$. Furthermore

$$
\dim \mathcal{T} = \dim \ker (H_{*_o})^{\perp} = 2m
$$

since, by Proposition 1.4, $F_{p,q}$ is an immersion. Introducing the $U(m)$-module

$$
\mathcal{R} = \text{span}\,\{\text{proj}_{\mathcal{H}^{p,q}_{\mathbf{C}P^m}}[Y]|Y \in \mathcal{T}\} \subset H^2_{p,q},
$$

by (1.5) and equivariance of $f_{p,q}$, we have

$$
(2.2) \qquad\qquad \mathcal{Z}_{p,q} = \text{span}\,(U(m+1)\cdot\mathcal{R}).
$$

LEMMA 2.3. *As $U(m)$-modules, we have*

$$
\mathcal{R} \otimes_{\mathbf{R}} \mathbf{C} \cong \mathcal{H}^{0,0}_{\mathbf{C}P^{m-1}} \oplus \mathcal{H}^{1,1}_{\mathbf{C}P^{m-1}}.
$$

122

PROOF: Denote by T_c the complex closure of T in $\mathcal{H}^{p,q}_{\mathbf{C}P^m}$. We can then write

$$\mathcal{R} = \operatorname{span}\{\operatorname{proj}_{T_c}[Y] | Y \in T\}.$$

As T_c is irreducible, it is either a complex or a totally real $U(m)$-submodule of $\mathcal{H}^{p,q}_{\mathbf{C}P^m}$.

CASE I: $T = T_c$.

We have

$$(2.3) \qquad \mathcal{H}^{p,q}_{\mathbf{C}P^m}|_{U(m)} = \sum_{\substack{0 \le r \le p \\ 0 \le s \le q}} \mathcal{H}^{r,s}_{\mathbf{C}P^{m-1}}$$

that follows by inspection or can be derived from the Branching Theorem of the next section as a particular case. (2.3) implies that $T \cong \mathcal{H}^{1,0}_{\mathbf{C}P^{m-1}}$ or $\mathcal{H}^{0,1}_{\mathbf{C}P^{m-1}}$ as complex $U(m)$-modules. Hence $\mathcal{R}$ consists of *all* Hermitian symmetric endomorphisms of T. Complexifying, we obtain

$$\mathcal{R} \otimes_{\mathbf{R}} \mathbf{C} \cong \operatorname{Hom}(T,T) \cong \mathcal{H}^{1,0}_{\mathbf{C}P^{m-1}} \otimes \mathcal{H}^{0,1}_{\mathbf{C}P^{m-1}} \cong \mathcal{H}^{0,0}_{\mathbf{C}P^{m-1}} \oplus \mathcal{H}^{1,1}_{\mathbf{C}P^{m-1}}.$$

CASE 2: $T \ne T_c$.

By (2.3), we have

$$T_c \cong \mathcal{H}^{1,0}_{\mathbf{C}P^{m-1}} \oplus \mathcal{H}^{0,1}_{\mathbf{C}P^{m-1}}.$$

(Recall that we assumed $q > 0$ so that $\mathcal{H}^{0,1}_{\mathbf{C}P^{m-1}}$ is actually a $U(m)$-component of $\mathcal{H}^{p,q}_{\mathbf{C}P^m}$.) Composition by the orthogonal projection $P : T_c \to \mathcal{H}^{1,0}_{\mathbf{C}P^{m-1}}$ induces a $U(m)$-module homomorphism

$$P^* : \mathcal{R} \to \operatorname{span}\{\operatorname{proj}_{\mathcal{H}^{1,0}_{\mathbf{C}P^{m-1}}}[Y] | Y \in \mathcal{H}^{1,0}_{\mathbf{C}P^{m-1}}\}.$$

(Note that P^* is well-defined as, for $Y \in T$, $\mathbf{C} \cdot Y \cap \mathcal{H}^{0,1}_{\mathbf{C}P^{m-1}} = \{0\}$ and so $\mathbf{C} \cdot Y$ projects down to a complex line in $\mathcal{H}^{1,0}_{\mathbf{C}P^{m-1}}$.) Easy computation shows that P^* is injective. It is also surjective as the range is the $U(m)$-module of all Hermitian symmetric endomorphisms of $\mathcal{H}^{1,0}_{\mathbf{C}Pm-1}$ (that splits into $\mathbf{R} \cdot I$ and the (irreducible) traceless part). As in Case 1, we obtain that $\mathcal{R} \otimes_{\mathbf{R}} \mathbf{C} \cong \mathcal{H}^{0,0}_{\mathbf{C}Pm-1} \oplus \mathcal{H}^{1,1}_{\mathbf{C}Pm-1}$. $\checkmark$

Now the cast for Proposition 2.1 is: $(G, K) = (U(m+1), U(m))$, $V = \mathcal{H}^{p,q}_{\mathbf{C}Pm}$ and $W_0 = \mathcal{R}$. By (2.2), W specializes to $\mathcal{Z}_{p,q}$ so that we arrive at:

THEOREM 2.4. Let $\bar{\mathcal{F}}_{p,q}$ denote the sum of those complex irreducible submodules of $\mathcal{H}^{p,q}_{\mathbf{C}Pm} \otimes \mathcal{H}^{q,p}_{\mathbf{C}Pm}$ that, when restricted to $U(m)$, do not contain $\mathcal{H}^{0,0}_{\mathbf{C}Pm-1}$ and $\mathcal{H}^{1,1}_{\mathbf{C}Pm-1}$. Then we have

$$\bar{\mathcal{F}}_{p,q} \subset \mathcal{F}_{p,q} \otimes_{\mathbf{R}} \mathbf{C}. \checkmark$$

Finally, we turn to the horizontal counterparts of $\mathcal{E}_{p,q}$ and $\mathcal{F}_{p,q}$. The cast for Proposition 2.1 is: $(G, K) = (U(m+1), U(m))$, $V = \mathcal{H}^{p,q}_{\mathbf{C}Pm}$ and

$$W_o = \text{span}\left\{(f_{p,q})_*(V_o) \cdot (f_{p,q})_*(X_o) | V_o \in \ker H_{*_o}, X_o \in (\ker H_{*_o})^{\perp}\right\}.$$

By equivariance of $f_{p,q}$, we have

$$\mathcal{A}_{p,q} = \text{span}\left\{U(m+1) \cdot ((f_{p,q})_*(V_o) \cdot (f_{p,q})_*(X_o)) | \right.$$
$$V_o \in \ker H_{*_o}, X_o \in (\ker H_{*_o})^{\perp}\}$$
$$= \text{span}\left\{U(m+1) \cdot W_o\right\}.$$

THEOREM 2.5. Let $\bar{\mathcal{E}}^*_{p,q}$ (resp. $\bar{\mathcal{F}}^*_{p,q}$) denote the sum of those complex irreducible submodules of $\bar{\mathcal{E}}_{p,q}$ (resp. $\bar{\mathcal{F}}_{p,q}$) that, when restricted to $U(m)$, do not contain $\mathcal{H}^{1,0}_{\mathbf{C}Pm-1}$ and $\mathcal{H}^{0,1}_{\mathbf{C}Pm-1}$. Then we have

$$\bar{\mathcal{E}}^*_{p,q} \subset \mathcal{E}^*_{p,q} \otimes_{\mathbf{R}} \mathbf{C} \ (\text{resp.}\, \bar{\mathcal{F}}^*_{p,q} \subset \mathcal{F}^*_{p,q} \otimes_{\mathbf{R}} \mathbf{C}).$$

PROOF: Notice that $W_o \cong \mathcal{T}$ as real $U(m)$-modules and hence

$$\mathcal{T} \otimes_{\mathbf{R}} \mathbf{C} \cong \mathcal{H}^{0,1}_{\mathbf{C}Pm-1} \oplus \mathcal{H}^{0,1}_{\mathbf{C}Pm-1}$$

(cf. the proof of Lemma 2.3). Both statements now follow from Proposition 2.1. $\checkmark$

§3. Decomposition of $\mathcal{H}_{\mathbf{C}Pm}^{p,q} \otimes \mathcal{H}_{\mathbf{C}Pm}^{q,p}$; the Littlewood-Richardson rule

The representation theory of the unitary group is simpler than that of the orthogonal group so that we recall only the bare minimum needed in the sequel. For more details, consult [Humphreys][Weyl][Robinson][Naimark-Stern]. Let $T = U(m+1)$ be the standard maximal torus

$$T = \{ \operatorname{diag}(e^{i\theta_1}, \ldots, e^{i\theta_{m+1}}) | \theta_j \in \mathbf{R}, j = 1, \ldots, m+1 \}$$

that contains the center $S^1 = \{ \operatorname{diag}(e^{i\theta}, \ldots, e^{i\theta}) | \theta \in \mathbf{R} \}$ of $U(m+1)$. Recall that the Weyl group $W_{U(m+1)}$ is nothing but the symmetric group $\mathcal{S}_{m+1}$ on $m+1$ letters acting on T by permuting the diagonal elements. As in the case of the orthogonal group, a complex irreducible $U(m+1)$-module V is determined by its highest weight ρ, an element of $\mathbf{Z}^{m+1}$. We write $V = V_{U(m+1)}^{\rho}$ up to isomorphism. Clearly

$$\rho_1 \geq \rho_2 \geq \cdots \geq \rho_{m+1},$$

where $\rho = (\rho_1, \rho_2, \ldots, \rho_{m+1}) \in \mathbf{Z}^{m+1}$. Note that the center $S^1 \subset U(m+1)$ acts on $V_{U(m+1)}^{\rho}$ by the single weight $\sum_{j=1}^{m+1} \rho_j$. In particular, we have

(3.1) $$\mathcal{H}_{\mathbf{C}Pm}^{p,q} \cong V_{U(m+1)}^{(p,0,\ldots,0,-q)}.$$

The Weyl dimension formula reads as

$$\dim_{\mathbf{C}} V_{U(m+1)}^{\rho} = \frac{\prod_{1 \leq r < s \leq m+1}(\rho_r - \rho_s - r + s)}{\prod_{1 \leq r < s \leq m+1}(-r + s)}.$$

The Branching Rule takes the form

(3.2) $$V_{U(m+1)}^{\rho}|_{U(m)} \cong \sum_{\sigma} V_{U(m)}^{\sigma},$$

where the summation runs over all $\sigma \in \mathbf{Z}^m$ for which

(3.3)
$$\rho_1 \geq \sigma_1 \geq \ldots \geq \rho_m \geq \sigma_m \geq \rho_{m+1}.$$

REMARK: Note that (2.3) follows from (3.2)-(3.3) via (3.1). These formulas also imply that every complex class 1 module with respect to $(U(m+1), U(m))$ is of the form $\mathcal{H}_{\mathbf{C}Pm}^{p,q}$ for some $p, q \geq 0$. (Compare with the remark before Theorem 2.2 in Chapter II.)

THEOREM 3.1. *Let* $p > q > 0$ *and* $m \geq 2$. *Then, for* $m \geq 3$, *we have*

$$\mathcal{H}_{\mathbf{C}Pm}^{p,q} \otimes \mathcal{H}_{\mathbf{C}Pm}^{q,p} \cong \sum_{b=0}^{a} \sum_{c=0}^{\min\{b,q,a-b\}} \sum_{d=0}^{\min\{b,q,e\}} \big[\min\{b-c, b-d, q-c, q-d,$$
(3.4)
$$b+c-2d, a-b-c\} + 1 \big] V^{(b,c,0,\ldots,0,-d,d-b-c)},$$

where $a = p + q$ *and* $e = |[\frac{b+c}{2}]|$. *For* $m = 2$, *we have*

(3.5)
$$\mathcal{H}_{\mathbf{C}P^2}^{p,q} \otimes \mathcal{H}_{\mathbf{C}P^2}^{q,p} \cong \sum_{b=0}^{a} [\min\{b, q, a-b\} + 1] \mathcal{H}_{\mathbf{C}P^2}^{b,b}$$
$$\oplus \sum_{c=1}^{q} \sum_{b=0}^{a-2c} [\min\{b, q-c, a-b-2c\} + 1]$$
$$\times \{V_{U(3)}^{(b+c,c,-b-2c)} \oplus V_{U(3)}^{(b+2c,-c,-b-c)}\}.$$

REMARK: (3.4) and (3.5) remain valid for $p = q$ as well. Nevertheless, to simplify the argument below, we assume $p > q$.

Before the proof we derive various consequences of these decompositions. First, every class 1 submodule of $\mathcal{H}_{\mathbf{C}Pm}^{p,q} \otimes \mathcal{H}_{\mathbf{C}Pm}^{q,p}$ with respect to $(U(m+1), U(m))$ is of the form $\mathcal{H}_{\mathbf{C}Pm}^{b,b}$ (since the center $S^1 \subset T$ acts on the tensor product trivially).

THEOREM 3.2. *Let* $p > q > 0$ *and* $m \geq 2$. *Then, we have*

(3.6)
$$\bar{\mathcal{E}}_{p,q} \otimes_{\mathbf{R}} \mathbf{C} \cong \mathcal{H}_{\mathbf{C}Pm}^{p,q} \otimes \mathcal{H}_{\mathbf{C}Pm}^{q,p} \Big/ \Big\{ \sum_{b=0}^{p+q} [\min\{b, q, p+q-b\} + 1] \mathcal{H}_{\mathbf{C}Pm}^{b,b} \Big\}$$

and

$$\bar{\mathcal{E}}^*_{p,q} \otimes_{\mathbf{R}} \mathbf{C} \cong \mathcal{H}^{p,q}_{\mathbf{C}P^m} \otimes \mathcal{H}^{q,p}_{\mathbf{C}P^m} / \left\{ \sum_{b=0}^{p+q} [\min\{b,q,p+q-b\}+1] \mathcal{H}^{b,b}_{\mathbf{C}P^m} \right.$$

(3.7)
$$\left. \oplus \sum_{b=1}^{p+q} \min\{b,q,p+q-b\} \{ V^{(b,1,0,\ldots,0,-b-1)}_{U(m+1)} \oplus V^{(b+1,0,\ldots,0,-1,-b)}_{U(m+1)} \} \right\}.$$

In particular, we have the lower estimates

$$\dim \mathcal{L}_{p,q} \geq \left[\binom{m+p}{p}\binom{m+q}{q} - \binom{m+p-1}{p-1}\binom{m+q-1}{q-1} \right]^2$$

(3.8)
$$- \sum_{b=0}^{p+q} [\min\{b,q,p+q-b\}+1] \left[\binom{m+b}{b}^2 - \binom{m+b-1}{b-1}^2 \right]$$

and

$$\dim \mathcal{L}^*_{p,q} \geq \left[\binom{m+p}{p}\binom{m+q}{q} - \binom{m+p-1}{p-1}\binom{m+q-1}{q-1} \right]^2$$

$$- \sum_{b=0}^{p+q} [\min\{b,q,p+q-b\}+1] \left[\binom{m+b}{b}^2 - \binom{m+b-1}{b-1}^2 \right]^2$$

(3.9)
$$- 2 \sum_{b=1}^{p+q} \min\{b,q,p+q-b\} \binom{m+b-1}{b+1} \frac{b(m+b+1)(m+2b+1)}{m(m-1)} \cdot \sqrt{}$$

REMARK: The lower bounds in (3.8)-(3.9) are obtained by applying the Weyl dimension formula to (3.6)-(3.7).

To get the decomposition of $\bar{\mathcal{F}}_{p,q} \otimes_{\mathbf{R}} \mathbf{C}$ and $\bar{\mathcal{F}}^*_{p,q} \otimes_{\mathbf{R}} \mathbf{C}$, we observe that the component

$$V^{(b,c,0,\ldots,0,-d,d-b-c)}_{U(m+1)}$$

does not contain $\mathcal{H}^{0,0}_{\mathbf{C}P^{m-1}}$ and $\mathcal{H}^{1,1}_{\mathbf{C}P^{m-1}}$ iff $c \geq 2$ or $d \geq 2$. This follows from the Branching Rule (3.2). Hence, using (3.4)-(3.5), Theorems 2.4 and 2.5 boil down to:

127

THEOREM 3.3. *Let $p > q > 0$ and $m \geq 2$. Then, we have*

$$\bar{\mathcal{F}}_{p,q} \otimes_{\mathbf{R}} \mathbf{C} \cong \bar{\mathcal{F}}^*_{p,q} \otimes_{\mathbf{R}} \mathbf{C} \cong \mathcal{H}^{p,q}_{\mathbf{C}P^m} \otimes \mathcal{H}^{q,p}_{\mathbf{C}P^m} / \left\{ \sum_{b=0}^{p+q} [\min\{b, q, p+q-b\} + 1] \mathcal{H}^{b,b}_{\mathbf{C}P^m} \right.$$

$$\oplus \sum_{b=1}^{p+q} \min\{b, q, p+q-b\} \{ V_{U(m+1)}^{(b,1,0,\ldots,0,-b-1)} \oplus V_{U(m+1)}^{(b+1,0,\ldots,0,-1,-b)} \}$$

$$\left. \oplus \sum_{b=1}^{p+q} \min\{b, q, p+q-b\} V_{U(m+1)}^{(b,1,0,\ldots,0,-1,-b)} \right\},$$

where, for $m = 2$, the last summation is absent. In particular, we have the lower estimate

$$\dim \mathcal{M}_{p,q} = \dim(\mathcal{F}_{p,q} \cap \mathcal{L}_{p,q}) \geq \left[\binom{m+p}{p} \binom{m+q}{q} - \binom{m+p-1}{p-1} \binom{m+q-1}{q-1} \right]^2$$

$$- \sum_{b=0}^{p+q} [\min\{b, q, p+q-b\} + 1] \binom{m+b-1}{b}^2 \frac{m+2b}{m}$$

$$- 2 \sum_{b=1}^{p+q} \min\{b, q, p+q-b\} \binom{m+b-1}{b+1}^2 \frac{b(m+b+1)(m+2b+1)}{m(m-1)}$$

$$- \sum_{b=1}^{p+q} \min\{b, q, p+q-b\} \binom{m+b-2}{b+1}^2 \frac{b^2(m+b)^2(m+2b)}{(m-1)^2 m},$$

where, again, for $m = 2$, the last summation is absent. $\checkmark$

PROBLEM: Prove a complex version of Theorem 4.6 of Chapter II about the number of geometrically distinct full eigenmaps of bidegree (p, q).

PROBLEM: Prove finiteness of the principal isotropy subgroup of the $U(m + 1)$-action on $\mathcal{L}_{p,q}$ (cf. Theorem 4.9 of Chapter II).

The rest of this section is devoted to the proof of Theorem 3.1. For $m = 2$ the computations are elementary and the decomposition in (3.5) follows from Steinberg's multiplicity formula (cf. [Humphreys]) as outlined in the following:

PROBLEM: Show the validity of the multiplicity formula

$$m\left[V_{U(3)}^{(b+c,c,-b-2c)} : \mathcal{H}^{p,q}_{\mathbf{C}P^2} \otimes \mathcal{H}^{q,p}_{\mathbf{C}P^2} \right] = \min\{b, q-c, p+q-b-2c\} + 1.$$

(Hint: By Steinberg's formula, this multiplicity is

$$\sum_{S,T \in W_{U(3)}} \det(ST) P\big[S(p+1,0,-q-1)+T(q+1,0,-p-1)+(-b-c-2,-c,b+2c+2)\big],$$

where P is the Kostant function (i.e. $P[\phi] = \#$ of distinct partitions of ϕ into the sum of positive roots (that are $(1,0,-1)$, $(1,-1,0)$, $(0,1,-1)$)). Since $p > q > 0$, the possible permutations for T are $(1)(2)(3)$, $(12)(3)$, $(1)(23)$, (123).)

In particular, we have

$$\mathcal{H}_{\mathbf{C}P^2}^{2,1} \otimes \mathcal{H}_{\mathbf{C}P^2}^{1,2} \cong \mathcal{H}_{\mathbf{C}P^2}^{0,0} \oplus 2\mathcal{H}_{\mathbf{C}P^2}^{1,1} \oplus 2\mathcal{H}_{\mathbf{C}P^2}^{2,2} \oplus \mathcal{H}_{\mathbf{C}P^2}^{3,3}$$
$$\oplus V_{U(3)}^{(1,1,-2)} \oplus V_{U(3)}^{(2,-1,-1)} \oplus V_{U(3)}^{(2,1,-3)} \oplus V_{U(3)}^{(3,-1,-2)}.$$

THEOREM 3.4. *For $m = 2$, the irreducible components of $\mathcal{E}_{2,1}$ are not class 1 with respect to $(U(3), U(2))$, i.e. we have*

$$\mathcal{E}_{2,1} \otimes_{\mathbf{R}} \mathbf{C} \cong V_{U(3)}^{(1,1,-2)} \oplus V_{U(3)}^{(2,-1,-1)} \oplus V_{U(3)}^{(2,1,-3)} \oplus V_{U(3)}^{(3,-1,-2)}.$$

In particular, the dimesion estimate (3.8) is sharp for $m = p = 2$ and $q = 1$.

PROOF: Assume that there exists a Hermitian symmetric endomorphism $C \in \mathcal{L}_{2,1}(\subset \mathcal{E}_{2,1})$ of $\mathcal{H}_{U(3)}^{2,1}$ which commutes with the subgroup $U(2) \subset U(3)$. By branching, we have

$$\mathcal{H}_{\mathbf{C}P^2}^{2,1}|_{U(2)} \cong \sum_{\substack{0 \leq r \leq 2 \\ 0 \leq s \leq 1}} \mathcal{H}_{\mathbf{C}P^1}^{r,s}$$

so that C is diagonal with respect to this decomposition. We now choose an orthonormal basis $\{f_{p,q}^i\}_{i=0}^{14}$ in $\mathcal{H}_{\mathbf{C}P^2}^{2,1}$ consisting of weight vectors of $U(2)$ in $\mathcal{H}_{\mathbf{C}P^1}^{r,s}$, $0 \leq$

$r \leq 2,\ 0 \leq s \leq 1$. In fact, using the coordinates $z, w, t \in \mathbf{C}$ we set

$$\mathcal{H}^{0,0}_{\mathbf{C}P^1} = \operatorname{span}\{(|z|^2 - |w|^2 - |t|^2)z\},$$

$$\mathcal{H}^{1,0}_{\mathbf{C}P^1} = \operatorname{span}\{1/\sqrt{2}(|w|^2 - 2|z|^2)w,\ 1/\sqrt{2}(|t|^2 - 2|z|^2)t\},$$

$$\mathcal{H}^{2,0}_{\mathbf{C}P^1} = \operatorname{span}\{\sqrt{3}w^2\bar{z},\ \sqrt{3}t^2\bar{z},\ \sqrt{2}\bar{z}wt\},$$

$$\mathcal{H}^{0,1}_{\mathbf{C}P^1} = \operatorname{span}\{\sqrt{3}z^2\bar{w},\ \sqrt{3}z^2\bar{t}\},$$

$$\mathcal{H}^{1,1}_{\mathbf{C}P^1} = \operatorname{span}\{(|w|^2 - |t|^2)z,\ \sqrt{2}z\bar{w}t,\ \sqrt{2}zw\bar{t}\},$$

$$\mathcal{H}^{2,1}_{\mathbf{C}P^1} = \operatorname{span}\{1/\sqrt{2}(|w|^2 - 2|t|^2)w,\ 1/\sqrt{2}(|t|^2 - 2|w|^2)t,\ \sqrt{3}w^2\bar{t},\ \sqrt{3}t^2\bar{w}\},$$

where the polynomials in the brackets are the components of $f_{2,1} : \mathbf{C}P^1 \to \mathbf{C}P^{14}$. The condition $C \in \mathcal{E}_{2,1}$ translates into $\langle C \circ f_{2,1},\ f_{2,1} \rangle = 0$. Expanding, $C = 0$ follows. $\checkmark$

REMARK: Similar result can be obtained for $\mathcal{H}^{3,1}_{\mathbf{C}P^2}$.

EXAMPLE 3.5: Using the explicit basis above, various (cubic) eigenmaps can be constructed between S^5 and S^{2n+1}, $6 \leq n \leq 14$. For example, $f : S^5 \to S^{13}$ is given by

$$f(z, w, t) = ((|z|^2 - \alpha|w|^2 - \beta|t|^2)z,\ (|w|^2 - \alpha|t|^2 - \beta|z|^2)w,$$

$$(|t|^2 - \alpha|z|^2 - \beta|w|^2)t,\ \gamma\bar{z}t^2,\ \gamma\bar{w}z^2,\ \gamma\bar{t}w^2,\ \delta\bar{z}wt),$$

where

$$\alpha = -1 + 2\sqrt{2},\ \beta = 3 - 2\sqrt{2},\ \gamma = 4\sqrt{-1 + 2\sqrt{2}},\ \delta = 2\sqrt{6}\sqrt{3 - 2\sqrt{2}}.$$

We now assume that $m \geq 3$. We first reformulate Theorem 3.1 as follows:

THEOREM 3.6. Let $p > q > 0$ and $m \geq 3$. Then, for $b \geq c \geq d > 0$, the multiplicity

$$(3.10) \qquad m\left[V^{(b,c,d,\dots)}_{U(m+1)} : \mathcal{H}^{p,q}_{\mathbf{C}P^m} \otimes \mathcal{H}^{q,p}_{\mathbf{C}P^m}\right] = 0.$$

Moreover, for $b_j \geq c_j \geq 0$, $j = 1, 2$, $b_1 + c_1 = b_2 + c_2$, we have

$$m\left[V_{U(m+1)}^{(b_1,c_1,0,\dots,0,-c_2,-b_2)} : \mathcal{H}_{\mathbb{C}P^m}^{p,q} \otimes \mathcal{H}_{\mathbb{C}P^m}^{q,p}\right]$$

$$= \min\{(b_1 + c_1)^+, (q - c_1)^+, (b_2 - c_2)^+, (q - c_2)^+,$$

(3.11) $\qquad (b_2 - c_1)^+, (b_1 - c_2)^+, (p + q - b_1 - c_1)^+\} + 1,$

where $^+$ denotes the positive part.

Once (3.10) and (3.11) are proved Theorem 3.1 follows easily. In fact, (3.10), along with self-duality of $\mathcal{H}_{\mathbb{C}P^m}^{p,q} \otimes \mathcal{H}_{\mathbb{C}P^m}^{q,p}$, implies that the only irreducible components of the tensor product are of the form $V_{U(m+1)}^{(b_1,c_1,0,\dots,-c_2,-b_2)}$. The center $S^1 \subset U(m + 1)$ acts trivially so that $b_1 + c_1 = b_2 + c_2$. The rest is a simple computation.

To determine the multiplicities we apply the Littlewood-Richardson rule (cf. [James] or [Robinson]). This we describe briefly as follows:

By Weyl's unitary trick (cf. [Weyl;2]), every irreducible tensor representation μ of rank k of the general linear group $GL(m + 1; \mathbf{C})$ restricts to an irreducible representation of the unitary group $U(m+1)$ and all (finite dimensional) representations of $U(m + 1)$ can be obtained in this way. Moreover, by Weyl's duality, to μ there corresponds a unique irreducible representation $[\mu]$ of the symmetric group $\mathcal{S}_k$ on k letters (cf. [Weyl;2]) (cf also §5 of Chapter II for the analogous theory of representations of the orthogonal group). $[\mu]$ determines (and is determined by) the Young tableau Σ_μ associated to μ. In what follows, we use the notation

$$\mu \leftrightarrow [\mu]$$

for duality. Given μ' and μ'', we have (with obviuos notation)

$$\mu' \otimes \mu'' \leftrightarrow \text{Ind}_{S_{k'} \times S_{k''}}^{S_{k'} + k''}([\mu'] \otimes [\mu''])$$

so that we have

$$m[\lambda : \mu' \otimes \mu''] = m\left[[\lambda] : \mathrm{Ind}_{S_{k'} \times S_{k''}}^{S_{k'+k''}}([\mu'] \otimes [\mu''])\right].$$

for the multiplicities. As a first step of the Littlewood-Richardson rule we add suitable elements of $\mathbf{Z}_+ \cdot (1, \ldots, 1)$ to the highest weight vectors of μ' and μ'' to make the components descend to zero. We also add the sum of these elements to the highest weight vector of λ. Each of the 3 vectors obtained in this way represents a Young tableau consisting of $m + 1$ rows; the coordinates representing the lengths of the respective rows. We superimpose the two largest tableaux and consider their difference called the complementary tableau. In the second step, we fill in the complementary tableau with the numbers $1, 2, \ldots, m$ and from each of these we use the amount given by the respective coordinate of the vector that has not been used so far (and which corresponds to the smallest tableau). The rules for filling are the following:

(1) In each row the numbers are nondecreasing;

(2) In each column the numbers are (strictly) increasing;

(3) When reading the sequence of numbers from right to left (a) the 1's are always O.K.;(b) given $i+1$ in the sequence, the number of previous i's is > then the number of previous $(i + 1)$'s.

The required multiplicity is the number of possible ways to fill in.

To prove (3.11) first, in this special case, the 3 vectors are as follows:

$$(b_1, c_1, 0, \ldots, 0, -c_2, -b_2) + (p + q, \ldots, p + q)$$

$$= (p + q + b_1, p + q + c_1, p + q, \ldots, p + q, p + q - c_2, p + q - b_2),$$

$$(p, 0, \ldots, 0, -q) + (q, \ldots, q) = (p + q, q, \ldots, q, 0),$$

$$(q, 0, \ldots, 0, -p) + (p, \ldots, p) = (p + q, p, \ldots, p, 0).$$

The two largest superimposed tableaux are:

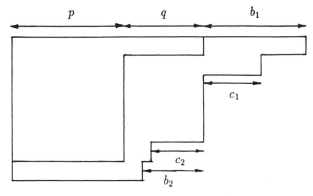

We denote by R_i, $i = 1, \ldots, m + 1$, the ith row of the complementary tableau. By (3) there can only be 1's in R_1. In particular, $b_1 \leq p + q$. By duality, we also have $b_2 \leq p + q$. By (1)-(3), the last c_1 entries of R_2 are filled with 2's. (Proof: There cannot be 1's there by (2). If there were an entry $i \geq 3$ then the same were true for the last entry of R_2 by (1). Now, apply (3) to that entry to get a contradiction.) There are at most q 2's, so that we obtain $c_1 \leq q$ (since otherwise the multiplicity is zero.) By duality, we also have $c_2 \leq q$, in particular, the smaller tableau is entirely contained in the larger tableau. By (1), there can only be 1's and 2's in R_2 the former preceding the latter.

From R_3 to R_{m-1} there are $m - 1$ column entries. Thus, by (2) in R_3 there can only be 2's, 3's and 4's. However 4 cannot occur in R_3 since in that case the last entry of R_3 were 4 and applying (3) we would get a contradiction. Thus, in R_3, there are only 2's and 3's. Below the 2's of R_2 there must be 3's, by (2).

We are interested in how many 3's are in R_3 under the 1's in R_2. We call these 'jumps' since the respective column in R_2 and R_3 will be $\binom{1}{3}$.

We claim that there are exactly c_1 jumps.

First of all, if there we less than c_1 jumps then we would run out of the 2's which are q in number. (We used up c_1 2's in the last c_1 entries of R_2 and, by (2) under the 2's in R_2 there must be 3's.) Secondly, if there were more than c_1 jumps then we would apply (3) to the 3 occurring in the very first jump. The number of 3's then would exceed the number of 2's; this is a contradiction.

As a byproduct we also obtained that we used up all 2's. We call the space occupied by the first c_1 3's in R_3 the *critical box*. We show below that any (allowed) location of the critical box determines the rest of the filling-ins.

In R_4 there can only be 3's and 4's by repeating the argument above. Below the 3's of R_3 there must be 4's and below the 2's of R_3 there must be 3's because otherwise the 4's in R_4 would exceed the 3's in R_3 violating (3).

This argument can be carried out till R_{m-1} is filled up. It also follows that we would use up all numbers between 2 and $m-2$.

We now consider R_{m+1} which, at this point, can only be filled up by 1's, $(m-1)$'s and m's.

We claim that there can only be 1's and m's in R_{m+1}.

Assume the contrary. If there are no m's is R_{m+1} then R_{m+1} is filled up by 1's and $(m-1)$'s. We then take the first m in R_m (which certainly exists) and apply (3) to get a contradiction. If there is an m in R_{m+1} we take the first one and apply again (3) to get a contradiction.

As a byproduct we also obtained that the $(m-1)$'s have to be used up in R_{m-1} and R_m. Thus, below the $(m-2)$'s in R_{m-1} there must be $(m-1)$'s of R_m and the filling-in is unique.

Summarizing, we showed that once the location of the critical box is fixed there is only one way to fill in.

To get the exact constraints on the location of the critical box, denote by d the distance of the critical box from the right wall (i.e. there are exactly $d + c_1$ 2's in R_2). The critical box occupies R_3 whose length is q so that $d \leq g - c_1$. There are at least $(q - b_2)^+$ $(m-1)$'s in R_m (by (2)) so that there are at least $(q - b_2)^+$ 2's in R_3. This means that $d \leq q - c_1 - (q - b_2)^+$. Taking into account the previous estimate, we can replace this by $d \leq b_2 - c_1$. In particular, $c_1 \leq b_2$ is a general constraint on the tableaux (i.e. otherwise the multiplicity is zero). By duality, we

134

also have $c_2 \leq b_1$. Summarizing, we obtain

$$d \leq \min\{b_1 - c_1, q - c_1, b_2 - c_1\}.$$

For the lower bound, first note that the critical box cannot occupy he the last $c_2 - 1$ entries of R_3 only because in that case the q $(m-1)$'s cannot be filled in R_{m-1} and R_m. Thus, $d \geq c_2 - c_1$. Moreover, in R_{m+1} there are $p + q - b_2$ places for the m's so that R_m ends with at least $(q - (p + q - b_2))^+ = (b_2 - p)^+$ m's. Thus, $d \geq c_2 - c_1 + (b_2 - p)^+$. By the previous estimate, we can replace this by $d \geq b_1 - p$. Summarizing, we obtain

$$d \geq \max\{c_2 - c_1, b_1 - p\}.$$

Comparing the upper and lower bounds, we obtain that $b_1 + c_1 \leq p + q$ is a constraint on the tableaux. Finally we obtain that the multiplicity minus one is the minimum of the following numbers:

$$b_1 - c_1, q - c_1, b_2 - c_1$$

and

$$b_1 - c_1 - (c_2 - c_1) = b_1 - c_2, \quad q - c_1 - (b_1 - p) = p + q - b_1 - c_1$$

$$b_1 - c_1 - (b_1 - p) = p - c_1, \quad b_2 - c_1 - (c_2 - c_1) = b_2 - c_2$$

$$q - c_1 - (c_2 - c_1) = q - c_2, \quad b_2 - c_1 - (b_1 - p) = p - c_2.$$

Two of the numbers do not contribute in the minimum as $p > q$. The multiplicity formula (3.11) follows.

To prove (3.10) we apply a similar argument. In the tableau corresponding to

$$m[V^{(b,c,d,\ldots)} : \mathcal{H}_{\mathbf{C}Pm}^{p,q} \otimes \mathcal{H}_{\mathbf{C}Pm}^{q,p}]$$

R_1 is filled up by 1's as before. The last c entries of R_2 are again filled up by 2's and there can only be 1's and 2's in R_2. Similarly, by (3), the last d entries of R_3

135

are filled up by 3's and there can only be 2's and 3's in R_3. Denote by J the number of jumps $\binom{1}{3}$ in R_2 and R_3. Then, by (3), we have $J \leq c - d$. On the other hand, $J \geq c$ since otherwise we would run out of the q 2's. This is a contradiction so that the multiplicity is zero. $\checkmark$

§4. Decomposition of $S^2(\mathcal{H}_{\mathbb{C}Pm}^{p,p})$

In this final (and most technical) section we give a complete decomposition of

$$S^2(\mathcal{H}_{\mathbb{C}Pm}^{p,p}) = S^2(\mathcal{H}_{\mathbb{C}Pm}^{p,p}(\mathbf{R})) \otimes_{\mathbf{R}} \mathbf{C}, \ p \geq 1,$$

into irreducible $U(m+1)$-components.

THEOREM 4.1. *For* $m \geq 3$, *we have*

$$S^2(\mathcal{H}^{p,p}) = \sum_{b=0}^{2p} \sum_{c=0}^{\min\{b,2p-b\}} \sum_{d=0}^{\min\{b,p,e\}} \frac{1}{2} \big[\min\{b-c, b-d, p-c, p-d,$$

$$(4.1) \quad b+c-2d, 2p-b-c\} + 1 + m_0(b,c,d) \big] V^{(b,c,0,\ldots,0,-d,d-b-c)},$$

where $e = |[\frac{b+c}{2}]|$ *and* $m_0(b,c,d)$, $b+c \leq 2p$, *is given as follows:*
$m_0(b,c,d) = 0$ *for* $b \not\equiv c \pmod 2$. *For* $b \equiv c \pmod 2$, *we have*

$$m_0(b,c,d) = \begin{cases} -1 & \text{if } b, d \text{ are odd and } m \equiv 1 \pmod 4 \\ \\ 1 & \text{otherwise.} \end{cases}$$

For $m = 2$, *we have*

$$S^2(\mathcal{H}_{\mathbb{C}P2}^{p,p}) \cong \sum_{b=0}^{2p} \frac{1}{2} \big[\min\{b, 2p-b\} + 1 + \frac{1}{2}(1 + (-1)^b) \big] \mathcal{H}_{\mathbb{C}P2}^{b,b}$$

$$(4.2) \qquad \oplus \sum_{c=1}^{p} \sum_{b=0}^{2p-2c} \frac{1}{2} \big[\min\{b, 2p-b-2c\} + 1 + m_0(b,c) \big]$$

$$\times \big(V_{U(3)}^{(b+c,c,-b-2c)} \oplus V_{U(3)}^{(b+2c,-c,-b-c)} \big),$$

where

$$
m_0(b, c) = \begin{cases} (-1)^c & \text{for } b \text{ even and } b \le 2p - 2c \\ \\ 0 & \text{otherwise.} \end{cases}
$$

By Theorem 3.8 of Chapter II we immediately obtain the following decomposition of $\mathcal{E}_{\lambda_p} \otimes_R \mathbf{C}$:

THEOREM 4.2. *For $m \ge 2$, we have*

$$
\mathcal{E}_{\lambda_p} \otimes_R \mathbf{C} \cong S^2(\mathcal{H}_{\mathbf{C}P^m}^{p,p}) / \left\{ \sum_{b=0}^{2p} [\min\{b, 2p - b\} + 1 + \tfrac{1}{2}(1 - (-1)^b)] \mathcal{H}_{\mathbf{C}P^m}^{b,b} \right\}.
$$

In particular, we have

$$
\dim \mathcal{L}_{\lambda_p} = \frac{1}{2} \left\{ \binom{m+p}{p}^2 - \binom{m+p-1}{p-1}^2 \right\} \left\{ \binom{m+p}{p}^2 - \binom{m+p-1}{p-1}^2 - 1 \right\}
$$

$$
- \sum_{b=0}^{2p} [\min\{b, 2p - b\} + 1 + \tfrac{1}{2}(1 - (-1)^b)]
$$

$$
\times \left\{ \binom{m+b}{b}^2 - \binom{m+b-1}{b-1}^2 \right\}.
$$

REMARK: For $m = 2$, (4.2) implies that $S^2(\mathcal{H}_{\mathbf{C}P^2}^{1,1})$ consists of class 1 modules with respect to $(U(3), U(2))$. This means that the moduli space $\mathcal{L}_{\lambda_1}$ is trivial. For $m \ge 3$, we have

$$
\mathcal{E}_{f_{\lambda_1}} \otimes_{\mathbf{R}} \mathbf{C} \cong V_{U(m+1)}^{(1,1,0,\dots,0,-2)} \oplus V_{U(m+1)}^{(2,0,\dots,0,-1,-1)} \oplus V_{U(m+1)}^{(1,1,0,\dots,0,-1,-1)},
$$

and the last term is present or not according as $m \not\equiv 1 \pmod 4$ or $m \equiv 1 \pmod 4$.

As for minimal immersions, using Theorem 3.9 of Chapter II, we obtain

THEOREM 4.3. *For $m \ge 2$, we have*

$$
\bar{\mathcal{F}}_{\lambda_p} \otimes_{\mathbf{R}} \mathbf{C} \cong S^2(\mathcal{H}_{\mathbf{C}P^m}^{p,p}) / \Bigg\{ \sum_{b=0}^{2p} \frac{1}{2} [\min\{b, 2p - b\} + 1 + \tfrac{1}{2}(1 + (-1)^b)] \mathcal{H}_{\mathbf{C}P^m}^{b,b}
$$

$$
\oplus \sum_{b=2}^{2p} \frac{1}{2} [\min\{b, 2p - b\} + m_0(b, 1, 0)] \{ V^{(b,1,0,\dots,0,-b-1)} \oplus V^{(b+1,0,\dots,0,-1,-b)} \}
$$

$$
\oplus \sum_{b=1}^{2p} \frac{1}{2} [\min\{b, 2p - b\} + m_0(b, 1, 1)] V^{(b,1,0,\dots,0,-1,-b)} \Bigg\},
$$

137

where, for $m = 2$, $m_0(b, 1, 0)$ has to be replaced by $m_0(b-1, 1)$ and the last summation is absent. In particular, we have the lower estimate

$$
\begin{aligned}
\dim \mathcal{M}_{\lambda_p} \geq {} & \frac{1}{2}\left\{ \binom{m+p}{p}^2 - \binom{m+p-1}{p-1}^2 \right\} \left\{ \binom{m+p}{p}^2 - \binom{m+p-1}{p-1}^2 - 1 \right\} \\
& - \sum_{b=0}^{2p} \frac{1}{2}\left[\min\{b, 2p-b\} + 1 + \frac{1}{2}(1 + (-1)^b) \right] \binom{m+b-1}{b}^2 \frac{m+2b}{m} \\
& - \sum_{b=2}^{2p} \left[\min\{b, 2p-b\} + m_0(b, 1, 0) \right] \binom{m+b-1}{b+1}^2 \frac{b(m+b+1)(m+2b+1)}{m(m-1)} \\
& - \sum_{b=1}^{2p} \frac{1}{2}\left[\min\{b, 2p-b\} + m_0(b, 1, 1) \right] \binom{m+b-2}{b+1}^2 \frac{b^2(m+b)^2(m+2b)}{(m-1)^2 m}
\end{aligned}
$$

where, for $m = 2$, the same rules apply for the evaluation of the lower bound. $\checkmark$

PROBLEM: Prove that, for $m \geq 2$, $\mathcal{M}_{\lambda_1} = \{0\}$, in particular, $\bar{\mathcal{F}}_{\lambda_1} = \mathcal{F}_{\lambda_1}$. (Hint: Use Problem * before Proposition 3.4 of Chapter II.)

Finally, we turn to the proof of Theorem 4.1. The argument to be presented here is based on [Robinson, Chapters 3 and 5] with appropriate corrections and modifications.

Recall that the tensor product $\mu' \otimes \mu''$ of the irreducible representations μ' and μ'' of $U(m+1)$ of rank k' and k'', respectively, corresponds, by Weyl's duality, to the induced representation

$$
\operatorname{Ind}_{S_{k'} \times S_{k''}}^{S_{k'} + k''} ([\mu'] \otimes [\mu'']).
$$

Turning to the symmetric square, let μ be an irreducible representation of $U(m+1)$ of rank k. Consider the semidirect product

$$
T_k = S_2 \triangleright (S_k \times S_k) \subset S_{2k},
$$

where S_2 acts on $S_k \times S_k$ by permuting the factors. Then, by Weyl's duality, we have

$$
S^2(\mu) \leftrightarrow \operatorname{Ind}_{T_k}^{S_{2k}} (1 \triangleright ([\mu] \otimes [\mu]))
$$

138

and

$$\wedge^2(\mu) \leftrightarrow \operatorname{Ind}_{T_k}^{S_{2k}}(\operatorname{sgn} \triangleright ([\mu] \otimes [\mu])),$$

where 1 and sgn denote the 1-dimensional trivial and sign representations of $\mathcal{S}_2$. In what follows, we give a method to determine the multiplicities

$$m[\lambda : S^2(\mu)]$$

and

$$m[\lambda : \wedge^2(\mu)].$$

By duality, the problem of computing these multiplicities is equivalent to that of determining

$$m\big[[\lambda] : \operatorname{Ind}_{T_k}^{S_{2k}}(\varepsilon \triangleright ([\mu] \otimes [\mu]))\big], \ \varepsilon = 1 \text{ or sgn}.$$

First note that the multiplicity

$$m\big[[\lambda] : \operatorname{Ind}_{S_k \times S_k}^{S_{2k}}([\mu] \otimes [\mu])\big]$$

can be computed by using the Littlewood-Richardson rule (as it corresponds to the multiplicity of λ in the tensor product $\mu \otimes \mu$). On the other hand, we have

$$\operatorname{Ind}_{S_k \times S_k}^{S_{2k}}([\mu] \otimes [\mu]) = \operatorname{Ind}_{T_k}^{S_{2k}}\big(\operatorname{Ind}_{S_k \times S_k}^{T_k}([\mu] \otimes [\mu])\big).$$

Using Frobenius Reciprocity

$$\operatorname{Ind}_{S_k \times S_k}^{T_k}([\mu] \otimes [\mu]) = 1 \triangleright ([\mu] \otimes [\mu]) \oplus \operatorname{sgn} \triangleright ([\mu] \otimes [\mu])$$

so that

$$m\big[[\lambda] : \operatorname{Ind}_{S_k \times S_k}^{S_{2k}}([\mu] \otimes [\mu])\big] =$$

$$(4.3) \quad = m\big[[\lambda] : \operatorname{Ind}_{T_k}^{S_{2k}}(1 \triangleright ([\mu] \otimes [\mu]))\big] + m\big[[\lambda] : \operatorname{Ind}_{T_k}^{S_{2k}}(\operatorname{sgn} \triangleright ([\mu] \otimes [\mu]))\big].$$

139

Secondly, we develop a formula for the difference of the multiplicities occuring in the right hand side of (4.3).

By Frobenius Reciprocity, we have

$$m\big[[\lambda] : \mathrm{Ind}_{T_k}^{S_{2k}}(\varepsilon \triangleright ([\mu] \otimes [\mu]))\big] = m\big[\varepsilon \triangleright ([\mu] \otimes [\mu]) : [\lambda]|_{T_k}\big].$$

We write

$$(4.4) \qquad [\lambda]|_{T_k} = \sum_{[\pi] \in \hat{T}_k} m\big[[\pi] : [\lambda]|_{T_k}\big] \cdot [\pi],$$

where $\hat{G}$, for a finite group G, denotes the set of irreducible representations of G. The irreducible components $[\pi]$ can only be one of the following types:

A. $[\pi] = \mathrm{Ind}_{S_k \times S_k}^{T_k}([\mu'] \otimes [\mu''])$, where $[\mu'] \ncong [\mu'']$,

B. $[\pi] = \varepsilon' \triangleright ([\mu'] \otimes [\mu'])$, where $[\mu'] \in \hat{S}_k$ and $\varepsilon' = 1$ or sgn.

We evaluate the characters of both sides of (4.4) on $t \triangleright (1, C)$, where $C \in S_k$ and $t \in S_2$ is the generator. With obvious notation, $t \triangleright (1, C) = (C, C)$ so that we obtain

$$(4.5) \qquad \chi_{[\lambda]}(C, C) = \sum_{[\pi] \in \hat{T}_k} m\big[[\pi] : [\lambda]|_{T_k}\big] \cdot \chi_{[\pi]}(t \triangleright (1, C)).$$

LEMMA 4.4. *For $[\pi]$ of type A, we have*

$$\chi_{[\pi]}(t \triangleright (1, C)) = 0.$$

PROOF: Since $\{1, t\}$ forms a complete set of representatives for the cosets of $S_k \times S_k$ in T_k, the character of the induced representation

$$\mathrm{Ind}_{S_k \times S_k}^{T_k}([\mu'] \otimes [\mu''])$$

evaluated on $a \in T_k$ is given by

$$\chi_{[\mu'] \otimes [\mu'']}(a) + \chi_{[\mu'] \otimes [\mu'']}(tat),$$

140

where the first term is present only for $a \in \mathcal{S}_k \times \mathcal{S}_k$ and the second only for $tat \in \mathcal{S}_k \times \mathcal{S}_k$. For $a = t \triangleright (1, C)$, we have $a, tat \notin \mathcal{S}_k \times \mathcal{S}_k$ and the lemma follows. $\checkmark$

For $[\pi]$ of type B, we have $\chi_{[\pi]}(t \triangleright (1, C)) = \varepsilon'(t)\chi_{[\mu']}(C)$ so that (4.5) rewrites as

$$\chi_{[\lambda]}(C, C) =$$
$$= \sum_{[\mu'] \in \hat{\mathcal{S}}_k} \left(m\big[[\lambda] : \mathrm{Ind}_{T_k}^{S_{2k}}(1 \triangleright ([\mu'] \otimes [\mu']))\big] - m\big[[\lambda] : \mathrm{Ind}_{T_k}^{S_{2k}}(\mathrm{sgn} \triangleright ([\mu'] \otimes [\mu']))\big] \right) \chi_{[\mu']}(C).$$

On the other hand, we have

$$\chi_{[\lambda]}(C, C) = \sum_{[\mu'] \in \hat{\mathcal{S}}_k} m_0(\lambda : \mu') \cdot \chi_{[\mu']}(C),$$

where a method to determine the coefficients of the expansion will be given below. Comparing the equal terms, we obtain

$$m_0(\lambda : \mu) =$$
$$(4.6) \quad = m\big[[\lambda] : \mathrm{Ind}_{T_k}^{S_{2k}}(1 \triangleright ([\mu] \otimes [\mu]))\big] - m\big[[\lambda] : \mathrm{Ind}_{T_k}^{S_{2k}}(\mathrm{sgn} \triangleright ([\mu] \otimes [\mu]))\big].$$

Combining (4.3) and (4.6) we arrive at the following formulas

$$m\big[[\lambda] : \mathrm{Ind}_{T_k}^{S_{2k}}(1 \triangleright ([\mu] \otimes [\mu]))\big] =$$
$$(4.7) \quad = \frac{1}{2}\left(m\big[[\lambda] : \mathrm{Ind}_{S_k \times S_k}^{S_{2k}}([\mu] \otimes [\mu])\big] + m_0(\lambda : \mu) \right)$$

and

$$m\big[[\lambda] : \mathrm{Ind}_{T_k}^{S_{2k}}(\mathrm{sgn} \triangleright ([\mu] \otimes [\mu]))\big] =$$
$$(4.8) \quad = \frac{1}{2}\left(m\big[[\lambda] : \mathrm{Ind}_{S_k \times S_k}^{S_{2k}}([\mu] \otimes [\mu])\big] - m_0(\lambda : \mu) \right).$$

Equivalently, we have

$$(4.9) \qquad m[\lambda : S^2(\mu)] = \frac{1}{2}\left\{ m[\lambda : \mu \otimes \mu] + m_0(\lambda : \mu) \right\}$$

141

and

(4.10) $$m[\lambda : \wedge^2(\mu)] = \frac{1}{2}\{m[\lambda : \mu \otimes \mu] - m_0(\lambda : \mu)\}.$$

(Note that (4.7) and (4.8) are the corrected versions of 4.5.3 of [Robinson]).

Finally, we determine $m_0(\lambda : \mu)$. Denote by Σ_μ the Young tableau corresponding to μ and Σ_λ the Young tableau corresponding to λ in $\mu \otimes \mu$.

In the first step we fill Σ_λ with horizontal and vertical dominoes. If a filling exists then we define $\epsilon(\lambda : \mu) = (-1)^v$, where v is the number of the vertical dominoes in the filling. If no filling is possible then we define $\epsilon(\lambda : \mu) = 0$.

In the second step we construct the *hook graph* $H(\lambda)$ of the Young tableau Σ_λ by replacing the entry in each box of the tableau by the hook length of the box. The *hook length* of a box occuring in the i-th row R_i and j-th column C_j of Σ_λ is equal to

$$r_i - i + c_j - j + 1,$$

where r_i is the number of boxes in R_i and c_j is the number of boxes in C_j. Next, for each row index i, we place the number $(r_i - i)(\mathrm{mod}\ 2) = 0$ or 1 at the end of R_i as a subscript. Then we form the Young tableau Σ_{λ^0} from Σ_λ by keeping those boxes of Σ_λ which have even hook lengths and which, furthermore, appear in rows with subscript 0. Similarly, Σ_{λ^1} is the Young tableau consisting of those boxes of Σ_λ which have even hook lengths and occur in rows with subscript 1. The Young tableaux Σ_{λ^0} and Σ_{λ^1} correspond to representations of symmetric groups S_{k_0} and S_{k_1}, where $k_0 + k_1 = k$. With these preparations, we finally have

$$m_0(\lambda : \mu) = \epsilon(\lambda : \mu) \cdot m\big[[\mu] : \mathrm{Ind}_{S_{k_0} \times S_{k_1}}^{S_k}([\lambda^0] \otimes [\lambda^1])\big],$$

where the multiplicity on the right hand side can again be determined by the Littlewood-Richardson rule.

We illustrate the method by computing the multiplicity

$$m[\mathcal{H}^{b,b} : S^2(\mathcal{H}^{p,p})].$$

Here $\mu = \mathcal{H}^{p,p}$ is given by its highest weight vector $(p, 0, \ldots, 0, -p) \in \mathbf{Z}^{m+1}$ so that Σ_μ is constructed from the vector

$$(p, 0, \ldots, 0, -p) + (p, \ldots, p) = (2p, p, \ldots, p, 0) \in \mathbf{Z}^{m+1}$$

whose coordinates are the row-lengths of Σ_μ. As $\lambda = \mathcal{H}^{b,b}$ is in the tensor product $\mathcal{H}^{p,p} \otimes \mathcal{H}^{p,p}$ the Young tableau for λ is constructed from the vector

$$(b, 0, \ldots, 0, -b) + (2p, \ldots, 2p) = (2p + b, 2p, \ldots, 2p, 2p - b) \in \mathbf{Z}^{m+1}.$$

In particular, $b \leq 2p$ since otherwise the multiplicity is zero.

We claim that

$$(4.11) \qquad m_0(\lambda : \mu) = \epsilon(\lambda : \mu) \cdot m(\lambda : \mu) = \frac{1}{2}(1 + (-1)^b), \ b \leq 2p.$$

We first determine $\epsilon(\lambda : \mu)$ and find that

$$\epsilon(\lambda : \mu) = \begin{cases} 1 & \text{if } b \text{ is even} \\ 0 & \text{if } b \text{ is odd and } m \text{ is even} \\ -1 & \text{if } b \text{ and } m \text{ are odd.} \end{cases}$$

Indeed, for b even, the whole tableau Σ_λ can be filled by horizontal dominoes so that $v = 0$ in this case. Now let b be odd. By filling in using horizontal dominoes from right to left the remaining part is a Young tableau determined by the vector

$$(3, 2, \ldots, 2, 1) \in \mathbf{Z}^{m+1}.$$

For m even, complete filling is impossible so that $\epsilon(\lambda : \mu) = 0$. (Note that (4.11) follows in this case.) For m odd, we use m vertical dominoes and 1 horizontal

143

domino for filling so that $\epsilon(\lambda : \mu) = (-1)^m = -1$. The formula for $\epsilon(\lambda : \mu)$ now follows.

In the second step we determine the hook graph of Σ_λ and arrive at the following

$m+2p+b$	$m+2b+1$	$m+2b-1$	$b+m$	b		1
$m+2p-1$	$m+b$	$m+b-2$	$m-1$			
$2p+1$	$b+2$	b	1			
$2p-b$	1					

Case 1. b and m are even.

In this case $R_2, R_4, \ldots, R_m$ receive zero subscript and the Young tableau Σ_{λ^0} is determined by the vector

$$(p, \ldots, p) \in \mathbf{Z}^{m/2}.$$

Similarly,

$$(p + b/2, p, \ldots, p, p - b/2) \in \mathbf{Z}^{m/2}$$

determines Σ_{λ^1}. To compute $m(\lambda : \mu)$ we use the Littlewood-Richardson rule and superimpose the Young tableaux Σ_μ and Σ_{λ^1} to obtain

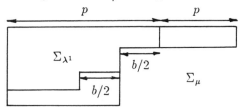

We have to fill the complementary tableau $\Sigma_\mu - \Sigma_{\lambda^1}$ by p 1's,..., p $m/2$'s, where the amount of each number is determined by the respective component of the vector $(p, \ldots, p)$ corresponding to Σ_{λ^0}. Clearly, there is only one way to fill in (observing the rules for filling) so that $m(\lambda : \mu) = 1$, and (4.11) follows in this case.

Case 2. b is even and m is odd.

144

In this case $R_2, R_4, \ldots, R_{m-1}, R_{m+1}$ receive zero subscript and Σ_{λ^0} is given by the vector

$$(p, \ldots, p, p - b/2) \in \mathbf{Z}^{(m+1)/2}.$$

Similarly, Σ_{λ^1} corresponds to

$$(p + b/2, p, \ldots, p) \in \mathbf{Z}^{(m+1)/2}.$$

We have to fill the complementary tableau $\Sigma_\mu - \Sigma_{\lambda^1}$ with p 1's,...,p $(m-1)/2$'s and $p - b/2$ $(m+1)/2$'s. Again, there is only one way so that $m(\lambda : \mu) = 1$ and (4.11) follows in this case.

Case 3. b and m are odd.

In this case $R_1, R_2, R_4, \ldots, R_{m-1}$ count in Σ_{λ^0} so that it corresponds to

$$(p + 1 + (b - 1)/2, p + 1, \ldots, p + 1) \in \mathbf{Z}^{(m+1)/2}.$$

Similarly, Σ_{λ^1} is given by

$$(p - 1, \ldots, p - 1, p - (b + 1)/2) \in \mathbf{Z}^{(m+1)/2}.$$

Clearly, $m(\lambda : \mu) = 0$ as, e.g. $p + 1$ 2's cannot fit in the complementary tableau. Thus (4.11) follows.

As a final step, note that the multiplicity formula

$$m[\mathcal{H}^{b,b} : \mathcal{H}^{p,p} \otimes \mathcal{H}^{p,p}] = \min\{b + 1, 2p - b + 1\}$$

has been established earlier. We obtain

$$m[\mathcal{H}^{b,b} : S^2(\mathcal{H}^{p,p})] = \frac{1}{2}\left(\min\{b + 1, 2p - b + 1\} + \frac{1}{2}(1 + (-1)^b)\right), \, b \leq 2p,$$

and

$$m[\mathcal{H}^{b,b} : \wedge^2(\mathcal{H}^{p,p})] = \frac{1}{2}\left(\min\{b + 1, 2p - b + 1\} - \frac{1}{2}(1 + (-1)^b)\right), \, b \leq 2p.$$

We now determine the complete decomposition of $S^2(\mathcal{H}^{p,p})$ and $\wedge^2(\mathcal{H}^{p,p})$. Setting $\mu = \mathcal{H}^{p,p}_{\mathbf{C}Pm}$ as above, this amounts to computing the multiplicities $m[\lambda : S^2(\mu)]$ and $m[\lambda : \wedge^2(\mu)]$ via (4.9) and (4.10). We determine $m_0(\lambda : \mu)$ for every irreducible component λ. We begin with the case $m = 2$ which is treated here separately.

145

PROPOSITION 4.5. *Assume that the highest weight vector of λ is*

$$(b+c,c,-b-2c)\,,\,b,c\geq 0.$$

Then, setting $m_0(\lambda,\mu) = m_0(b,c)$, we have $m_0(b,c) = (-1)^c$ for b even and $b \leq 2p - 2c$. Otherwise $m_0(b,c) = 0$.

PROOF: The Young tableau Σ_μ is determined by the vector

$$(p,0,-p)+(p,p,p)=(2p,p,0)\in \mathbf{Z}^3,$$

so that Σ_λ corresponds to

$$(2p+b+c,2p+c,2p-b-2c)\in \mathbf{Z}^3.$$

Splitting into cases according to the parities of b and c the result follows easily. $\checkmark$

Decomposing $\mathcal{H}^{p,p}_{\mathbf{C}P^m} \otimes \mathcal{H}^{p,p}_{\mathbf{C}P^m}$, we obtain

$$S^2(\mathcal{H}^{p,p}) = \sum_{b=0}^{2p} \frac{1}{2}\left[\min\{b+1,2p-b+1\}+\frac{1}{2}(1+(-1)^b)\right]\mathcal{H}^{b,b}$$

$$\oplus \sum_{c=1}^{p} \sum_{b=0}^{2p-2c} \frac{1}{2}\left[\min\{b+1,2p-b-2c+1\}+m_0(b,c)\right] \cdot$$

$$\times \left(V^{(b+c,c,-b-2c)} \oplus V^{(b+2c,-c,-b-c)}\right)$$

and

$$\wedge^2(\mathcal{H}^{p,p}) = \sum_{b=0}^{2p} \frac{1}{2}\left[\min\{b+1,2p-b+1\}-\frac{1}{2}(1+(-1)^b)\right]\mathcal{H}^{b,b}$$

$$\oplus \sum_{c=1}^{p} \sum_{b=0}^{2p-2c} \frac{1}{2}\left[\min\{b+1,2p-b-2c+1\}-m_0(b,c)\right] \cdot$$

$$\times \left(V^{(b+c,c,-b-2c)} \oplus V^{(b+2c,-c,-b-c)}\right).$$

REMARK: We have $m_0(b,0) = \frac{1}{2}(1+(-1)^b)$, $b \leq 2p$, in agreement with the earlier computations.

146

We now turn to the case $m \geq 3$. The decomposition formula for the tensor product $\mathcal{H}^{p,p}_{\mathbb{C}P^m} \otimes \mathcal{H}^{p,p}_{\mathbb{C}P^m}$ says that every irreducible component is of the form

$$\lambda = V^{(b_1, c_1, 0, \ldots, 0, -c_2, -b_2)},$$

where $b_j \geq c_j \geq 0$, $j = 1, 2$, and $b_1 + c_1 = b_2 + c_2$. Moreover, we have

$$m[V^{(b_1, c_1, 0, \ldots, 0, -c_2, -b_2)} : \mathcal{H}^{p,p} \otimes \mathcal{H}^{p,p}] =$$

$$= \min\{(b_1 - c_1)^+, (p - c_1)^+, (b_2 - c_2)^+, (p - c_2)^+,$$

$$(b_2 - c_1)^+, (b_1 - c_2)^+, (2p - b_1 - c_1)^+\} + 1,$$

where $^+$ means 'positive part'. Again it remains to compute $m_0(\lambda : \mu) = \epsilon(\lambda : \mu) m(\lambda : \mu)$.

PROPOSITION 4.6. $m_0(\lambda : \mu) = 0$ for $b_1 + c_1 = b_2 + c_2$ odd.

PROOF: We assume that $m \geq 4$ as $m = 3$ can be treated separately using the same ideas and somewhat simpler argument as below. The Young tableau Σ_μ is determined by the vector

$$(2p, p, \ldots, p, 0) \in \mathbf{Z}^{m+1}$$

so that Σ_λ corresponds to

$$(2p + b_1, 2p + c_1, 2p, \ldots, 2p, 2p - c_2, 2p - b_2) \in \mathbf{Z}^{m+1}.$$

Let R_i denote the ith row of Σ_λ. As $b_1 + c_1$ is odd R_1 and R_2 have the same subscript so that their boxes with even hook lengths end up in the firs two rows of λ^l, where $l \equiv b_1 + 1 \pmod{2}$. We distinguish four cases.

Case 1. b_1 and b_2 are even.

In this case c_1 and c_2 are odd and $l = 1$. Counting the number of even hook lengths in R_1 and R_2 we obtain that the first two rows of λ^1 have lengths $p + b_1/2$ and $p + (c_1 + 1)/2$ so that λ^1 does not fit in Σ_μ. Thus λ^0 has to fit in Σ_μ and we have

147

to use $p+(c_1+1)/2\,(>p)$ 2's in the complementary boxes. However, this leads to $m(\lambda:\mu)=0$ since, by the Littlewood-Richardson rule the first row has to be filled up by 1's and there cannot be more than one 2's in a single column.

Case 2. b_1 and b_2 are odd.

In this case c_1 and c_2 are even and $l=0$. The first two rows of λ^0 have lengths $p+(b_1+1)/2$ and $p+c_1/2+1$ so that the argument used above applies here yielding $m(\lambda:\mu)=0$.

Case 3. b_1 is even and b_2 is odd.

In this case c_1 is odd and c_2 is even and $l=1$. As $m\geq 4$, R_3 also receives the subscript 1. Counting the number of even hook lengths in R_3 we obtain that the length of the third row in λ^1 is $p+1$ or $p-1$ according as m is even or odd. The argument used in Case 1 applies here yielding that m is odd. We now calculate the total amount of boxes with even hook lengths in the hook graph which turns out to be $(m+1)p-4$. However the total amount of boxes in Σ_μ is $(m+1)p$ so that $m(\lambda:\mu)=0$ follows.

Case 4. b_1 is odd and b_2 is even.

This case is analogous to the previous one. Note that one can also reduce this to Case 3 based on duality which interchanges b_1 and b_2 (along with c_1 and c_2).$\sqrt{}$

THEOREM 4.7. For $b_1+c_1=b_2+c_2$ even and $\leq 2p$, we have

$$
m_0(\lambda:\mu)=\begin{cases} -1 & \text{if } b_1,b_2 \text{ are odd and } m\equiv 1 (\mathrm{mod}\, 4) \\[2em] 1 & \text{otherwise.} \end{cases}
$$

In all other cases $m_0(\lambda:\mu)=0$.

PROOF: By filling Σ_λ with dominoes it follows easily that

$$\epsilon(\lambda : \mu) = \begin{cases} -1 & \text{if } b_1, b_2 \text{ are odd and } m \equiv 1 (\mathrm{mod}\, 4) \\ \\ 1 & \text{otherwise.} \end{cases}$$

We claim that

$$m(\lambda : \mu) = \begin{cases} 1 & \text{if } b_1 + c_1 \le 2p \\ \\ 0 & \text{otherwise.} \end{cases}$$

This can be done by case-by-case verification separating according to the parities of b_1, b_2 and m. We give the details in the case when b_1, b_2 and m are even.

In Σ_λ, the rows $R_2, R_4, \ldots, R_{m-2}$ and R_m receive subscript 0. Working out the hook lengths we obtain that the lengths of the rows of λ^0 are given by the vector

$$(p + c_1/2, p, \ldots, p, p - c_2/2) \in \mathbf{Z}^{m/2}.$$

Similarly, λ^1 is determined by the vector

$$(p + b_1/2, p, \ldots, p, p - b_2/2) \in \mathbf{Z}^{m/2+1}.$$

Superimposing, we have to fill the complementary tableau $\Sigma_\mu - \Sigma_{\lambda^1}$ by $p + c_1/2$ 1's, p 2's, ..., p $(m/2 - 1)$'s and $p - c_2/2$ $(m/2)$'s. As the first row of the skew tableau has to be filled with 1's, we have

$$p + c_1/2 \le p - b_1/2 + p$$

since otherwise $m(\lambda : \mu)$ is zero. (Any column contains distinct entries.) This translates into

$$c_1 + b_1 \le 2p.$$

Assuming this, it is easy to see that there is only one way to fill in the skew tableau so that $m(\lambda : \mu) = 1$. $\checkmark$

Bibliography

[Barbasch D.-Glazebrook J.F.-Toth G.] *Harmonic maps between complex projective spaces*, Geometriae Dedicata (to appear).

[Berger M.] *Geometry I-II*, Springer, 1987.

[Berger M.-Gauduchon P.-Mazet E.] *Le Spectre d'une variété Riemannienne*, LNM 194, 1971.

[Börner H.] *Representations of groups*, North Holland, Amsterdam, 1963.

[Bredon G.] *Introduction to compact transformation groups*, Academic Press, New York, 1972.

[Calabi E.] *Minimal immersions of surfaces in Euclidean spheres*, J. Diff. Geom., 1 (1967) 111-125.

[Cartan É.] 1. *Sur la détérmination d'un système orthogonal complet dans un espace de Riemann symétrique clos*, Rend. Circ. Mat. Palermo, 53 (1929) 217-252. 2. *Sur quelques familles remarquables d'hypersurfaces*, C.R. Congrès Math. Liège (1939) 30-41.

[DoCarmo M.-Wallach N.] *Minimal immersions of spheres into spheres*, Ann. of Math., 93 (1971) 43-62.

[Eells J.-Lemaire L.] 1. *A report on harmonic maps*, Bull. London Math. Soc., 10 (1978) 1-68. 2. *Selected topics in harmonic maps*, Reg. Conf. Ser. in Math., No. 50, AMS, 1982. 3. *Another report on harmonic maps*, Bull. London Math. Soc., 20 (1988) 385-524.

[Eells J.-Sampson J.H.] *Harmonic mappings of Riemannian manifolds*, Amer. J. Math., 86 (1964) 109-160.

[Eells J.-Wood J.C.] *Harmonic maps from surfaces to complex projective spaces*, Adv. in Math., Vol.49, No.3 (1983) 217-263.

[Fuller F.B.] *Harmonic mappings*, Proc. Nat. Acad. Sci., 40 (1954) 987-991.

[Hsiang Wu-Yi] 1. *On the compact, homogeneous minimal submanifolds*, Proc. Nat. Acad. Sci. U.S.A., 56 (1966) 5-6. 2. *Cohomology theory of topological transformation groups*, Springer, 1975.

[Humphreys J.E.] *Introduction to Lie algebras and representation theory*, Springer, 1980.

[Husemoller D.] *Fibre bundles*, McGraw-Hill, New York, 1966.

[James G.D.] *The representation theory of the symmetric group*, LNM 682, Springer, New York, 1978.

[Lawson H.B.] *Lectures on minimal submanifolds*, IMPA, 1970.

[Mashimo K.] 1. *Degree of the standard isometric minimal immersions of the symmetric spaces of rank one into spheres*, Tsukuba J. Math. Vol.5, No.2 (1981) 291-297. 2. *Order of the standard isometric minimal immersions of cross as helical geodesic immersions*, Tsukuba J. Math. Vol.7, No.2 (1983) 257-263.

[Mostow G.D.] *Equivariant embeddings of Euclidean space*, Ann. of Math., 65 (1957) 432-446.

[Muto Y.] *The space W_2 of isometric minimal immersions of the three-dimensional sphere into spheres*, Tokyo J. Math. Vol.7, No.2 (1984) 337-358.

[Münzner F.] *Isoparametrische Hyperflächen in Sphären*, Math. Ann. 251 (1980) 57-71.

[Naimark M.A.-Stern A.I.] *Theory of group representations*, Springer,1982.

[Parker M.] *Orthogonal multiplications in small dimensions*, Bull. London Math. Soc. 15 (1983) 368-372.

[Parks J.S.-Urakawa H.] *Classification of harmonic mappings of constant energy density into spheres*, Preprint (1989).

[Robinson G.deB.] *Representation of the symmetric group*, University of Toronto Press (1961).

[Sakamoto K.] *Helical immersions into a unit sphere*, Math. Ann., 261 (1982) 63-80.

[Smith R.T.] *Harmonic mappings of spheres*, Thesis, Warwick University, 1972.

[Takahashi T.] *Minimal immersions of Riemannian manifolds*, J. Math. Soc. Japan, 18 (1966) 380-385.

[Toth G.] 1. *On rigidity of harmonic mappings into spheres*, J. London Math. Soc. (2) (1982) 475-486. 2. *Classification of quadratic harmonic maps of S^3 into spheres*, Indiana Univ. Math. J. Vol.36, No.2 (1987) 231-239. 3. *Harmonic and minimal maps*, E. Horwood Series, Halsted Press, John Wiley and Sons, 1984.

[Tsukada K.] *Helical geodesic immersions of compact rank one symmetric spaces* (to appear).

[Urakawa H.] 1. *Stability of harmonic maps and eigenvalues of the Laplacian*, Trans. Amer. Math. Soc. 301 (1987) 557-589. 2. *Minimal immersions of projective spaces into spheres*, Tsukuba J. Math. 9, No.2. (1985) 321-347.

[Vilekin N.I.] *Special functions and the theory of group representations*, Amer. Math. Soc. Translations of Mathematical Monographs, Vol.22, 1968.

[Xin Y.L.] *Some results on stable harmonic maps*, Duke Math. J. 47 (1980) 609-613.

[Wallach N.R.] *Minimal immersions of symmetric spaces into spheres*, in Symmetric Spaces, Dekker, New York (1972) 1-40.

[Weyl H.] 1. *Classical groups*, Princeton Mathematical Series, 1. Princeton Press, 1946. 2. *The theory of groups and quantum mechanics*, Dover Publications, 1950.

[Wood R.] *Polynomial maps from spheres to spheres*, Inventiones Math., 5 (1968) 163-168.

[Zhelobenko D.P.] *Compact Lie groups and their representations*, Amer. Math. Soc., Providence, Rhode Island, 1973.

Perspectives in Mathematics